FOR THE SAKE OF
FORESTS AND GODS

FOR THE SAKE OF FORESTS AND GODS

Governing Life and Livelihood in the Philippine Uplands

Wolfram H. Dressler

SOUTHEAST ASIA PROGRAM PUBLICATIONS

an imprint of

CORNELL UNIVERSITY PRESS

Ithaca and London

First published 2025 by Cornell University Press

Library of Congress Cataloging-in-Publication Data
Names: Dressler, Wolfram Heinz, author.
Title: For the sake of forests and gods : governing life and livelihood in the
    Philippine uplands / Wolfram H. Dressler.
Description: Ithaca : Cornell University Press, 2025. | Includes bibliographical
    references and index.
Identifiers: LCCN 2024011505 (print) | LCCN 2024011506 (ebook) |
    ISBN 9781501779251 (hardcover) | ISBN 9781501779268 (paperback) |
    ISBN 9781501779275 (epub) | ISBN 9781501779282 (pdf)
Subjects: LCSH: Palawan (Philippine people)—Social conditions—21st century. |
    Palawan (Philippine people)—Economic conditions—21st century. |
    Non-governmental organizations—Philippines—Influence. |
    Evangelicalism—Philippines—Influence.
Classification: LCC GN671.P5 D74 2025 (print) | LCC GN671.P5 (ebook) |
    DDC 305.899/21—dc23/eng/20240712
LC record available at https://lccn.loc.gov/2024011505
LC ebook record available at https://lccn.loc.gov/2024011506

For the Pala'wan

In loving memory of Burkhard Dressler

Contents

Acknowledgments

For over two decades, every journey into the uplands of Palawan has been embraced by the warm hospitality, deep humility, and vivid stories of Indigenous highlanders—whether among the Pala'wan, Tagbanua, or Batak peoples. I hope that the pages of this book resonate with your stories of struggle and hope and positively influence the changes unfolding around you. Thank you for opening your homes to me, sharing stories, and indulging my awkward pronunciations and gestures. This book is dedicated to you.

This book would not have been possible without the support of my dearest of friends, Ched Limsa. Your advocacy, kindness, and patience are unmatched. I also thank Jessie Varquez, Rogelio Rodrigo, Ana Bibal, and Michael Pido for supporting my research at key moments; it was done in solidarity with you and the many nongovernmental organizations who work *with* (rather than on) Indigenous highlanders in defense of ancestral lands. You know who you are.

Various family members, friends, and colleagues have been crucial in supporting me emotionally and intellectually over the five years it took to put this book together. My parents, Barbara and Burkhard Dressler, always listened attentively, asking probing questions about the work I do. Upon writing the last chapter, I lost my beloved father, Burkhard Dressler, to whom I dedicate this book. My cherished partner, Ilka Barr, has been a steadfast pillar of support and kindness. Thank you for dealing with all the emotional ups and downs along the way. No more books for a while.

Numerous colleagues have read and commented on different sections and drafts. Bram Büscher, Rob Fletcher, Sian Sullivan, Michael Fabinyi, Andrea Procter, Jessica Clendenning, and Ilan Wiezel all provided critical feedback, for which the book is hopefully much improved. Over the years, Charles Macdonald, Nicole Revel, Norli Collili, Noah Theriault, Will Smith, and Dario Novellino have answered many of my questions concerning Pala'wan culture, language, social organization, and livelihood. A special thanks to Jacob Henry for his meticulous copyediting and to Chandra Jayasuriya for her fantastic mapmaking skills. I am also very thankful for the support and attention to detail from CUP editor Sarah Grossman. You have all shaped the book's content in different ways. Any errors and omissions are my own.

The Australian Research Council's Future Fellowship scheme and Discovery Project grants funded much of this research, allowing for multiple trips to Palawan during the pressures of teaching, administration, and graduate supervision.

Abbreviations

ADB	Asian Development Bank
ADSDPP	Ancestral Domain Sustainable Development and Protection Plan
AFM	Adventist Frontier Missions
BARMM	Bangsamoro Autonomous Region in Muslim Mindanao
CADC/Ts	Certificates of Ancestral Domain Claim/Titles
CCA	Community Conservation Agreement
CI	Conservation International
CSSC	Commission on Social and Special Concerns (formerly known as "Indigenous Peoples' Apostolate")
DAO 2	Departmental Administrative Order No. 2
DENR	Department of Environment and Natural Resources
ECAN	Environmentally Critical Areas Network
ELAC	Environmental Legal Assistance Center
FAR	Family Approach to Reforestation
FLUPs	forest land-use plans
FOMP	Forest Occupancy Management Program
IEC	information-education-communication
Ip	Indigenous peoples
IPA	Indigenous Peoples' Apostolate (now the Commission on Social and Special Concerns)
IPRA	Indigenous Peoples' Right Act
ISFP	Integrated Social Forestry Program
LAPS	Livelihood Appraisal and Product Scanning
LGU	local government unit
m asl	meters above sea level
MILF	Moro Islamic Liberation Front
MMPL	Mount Mantalingahan Protected Landscape
NATRIPAL	United Tribes of Palawan (Nagkakaisang mga Tribu ng Palawan)
NCIP	National Commission of Indigenous Peoples
NTFP	nontimber forest product
PANLIPI	Legal Assistance Center for Indigenous Filipinos (Tanggapang Panligal ng Katutubong Pilipino)
PCSD	Palawan Council for Sustainable Development
PD	Presidential Decree

PES	payment for ecosystem services
PIADP	Palawan Integrated Development Project
PNNI	Palawan NGO Network Incorporated
POs	people's organizations
PPC	Puerto Princesa City
PTFPP	Palawan Tropical Forest Protection Programme
SDA	Seventh-day Adventist
SEP	Strategic Environmental Plan
SPPC	South Palawan Planning Council
USAID	U.S. Agency for International Development

Note on Language

Scholars spell the demonym "Pala'wan" in various ways, from "Palawan," "Pala-wan," "Pelawan," "Pälawan," and "Palawanän," among others, depending on the dialectic of Pala'wan subgroups (Macdonald 2007; Theriault 2017; Smith 2021; Iskander 2021). In this book, I use the spelling "Pala'wan" to differentiate the people from the island name, Palawan. In line with Macdonald, I use the phoneme /e/, sometimes transcribed as /ä/, to stress the open back vowel, as in the "a" in English "ball" (Macdonald 2007, xiii). However, I only do so in reference to significant ritual, ceremony, livelihood practices and place names. I use "et" (meaning "of") in full in formal descriptive writing (e.g., Empu *et* parey) and then the contraction (elision) of "et" as "'t" in vernacular descriptions or quotes from Pala'wan (such that phrasing like "Kundu et Tinkep" appears as "Kundu 't Tingkep.") (Thank you to Dario Novellino, Noah Theriault and Will Smith for further insights into the matter.) Most discussions and interviews were conducted in Tagalog/Filipino (the lingua franca), which many Pala'wan speak to a greater or lesser extent (e.g., because of daily lowland interactions, such as wage labor and trading. Farther in the interior, Pala'wan tend to speak only their language). I am conversant in Tagalog but have learned only key Pala'wan words and phrases. I therefore relied heavily on a local Pala'wan translator when interviews and discussions switched between Tagalog and Pala'wan. In cases when Pala'wan were reluctant to speak Tagalog or had difficulty understanding the language, interviews and discussions were done in the Pala'wan language and translated accordingly. As often happens, these discussions necessarily unfolded slowly with repeat explanations and careful notetaking on my part. While some meaning was inevitably lost in translation and transcriptions, I deliberately only draw on quotes and meanings that arose frequently and could be substantiated through triangulation. Important words and their meanings were also verified by Charles Macdonald's online Pala'wan dictionary and in consultation with him and others who speak Pala'wan.

FOR THE SAKE OF
FORESTS AND GODS

INTRODUCTION
Governing the Ungoverned

After walking for several hours along a narrow forest trail, I reached the Pala'wan hamlet (*sitio*) known as Kamantian, in the interior of the Mount Mantalingahan Range in southern Palawan, the Philippines (see map I.1). The hamlet sat, nestled high in the mountains, within a patchwork of older and recently cleared swiddens. In 2013, I visited this and other upland areas to learn more about how environmental decline and restrictive conservation were affecting the livelihoods of the Indigenous Pala'wan—an autochthonous people whose swiddens, hunting, and harvesting have drawn heavily on this highland forest landscape for centuries (Macdonald 2007). In 2009, the Philippine state joined with the European Union and Conservation International (CI) to establish the 120,000-hectare Mount Mantalingahan Protected Landscape (MMPL) area, which incorporated large swaths of Pala'wan ancestral territory. The protected landscape, now a global biodiversity "hotspot," soon limited Pala'wan forest access through newly designated "sustainable use" and "no use" zoning regimes. Shortly thereafter, the area's endemic species—Palawan pangolins, bearded wild boar, and the Palawan hornbill—appeared in *National Geographic* (Priit 2002) and were declared in need of protection under CI's "shared goal" of "no more forest loss" and "zero extinction" (Conservation International 2021).[1]

CI's goals of conserving old-growth forests and achieving zero extinction implicated the roughly fifty thousand Pala'wan people living across 145 kilometers of rugged forest terrain (Macdonald 2007). They would need to convert their so-called slash and burn (swidden) agriculture into permanent, fixed-plot agriculture and stop hunting forest animals as they had for generations. I was initially more familiar with how fellow *katutubo* (Indigenous) Tagbanua and

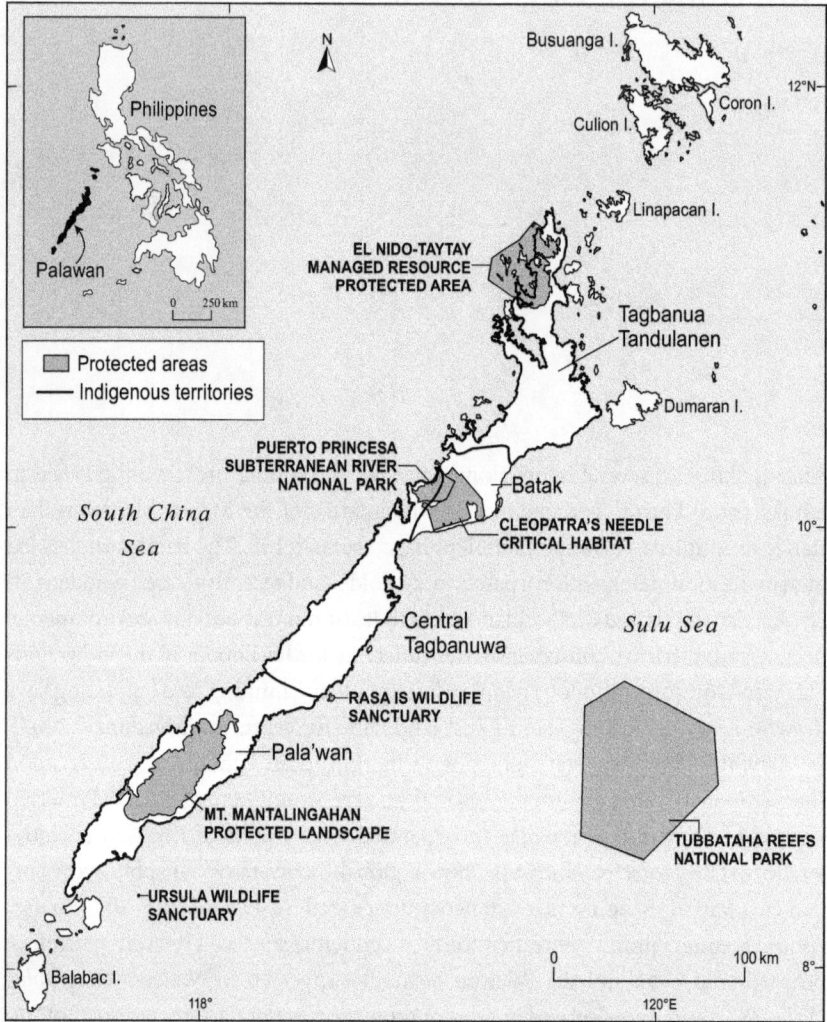

MAP I.1. Indigenous territories and protected areas on Palawan. Cartography © Chandra Jayasuriya, 2022.

Batak contended with coercive conservation in the north (Dressler 2009) and naively thought that the Pala'wan had not been subjected to the same reforms. As it turned out, these same biological and political reforms—modern, intensive ("deforestation-free") agriculture, and low-hunting diets—had long been imposed upon the area amid livelihood decline and risk.

Pala'wan villagers explained that their primary sources of subsistence, swidden plots and fallows, were in relative decline. Soils were no longer "fat" (*mataba*); rice yields were low; surplus was nearly absent; and, with establishment of the new protected area, some Pala'wan worried about bans against swidden (*uma*). The local *panglima* Kulibit with whom I had first discussed my research, lamented that "without our uma, there is no food. Clearing in older forest is better, but now we're afraid because of rangers, so there's less rice in our fields."[2] Upland peoples now faced multiple overlapping constraints, including oversight by more organizations with complex rules, land and forest enclosures, shorter fallows, and less game (Dressler 2009; Montefrio 2017). Other Pala'wan suggested that a regular cash income was now needed to purchase additional rice and other household goods to meet changing family needs and aspirations. Growing financial constraints required that they sell more nontimber forest products (NTFPs) and frequently labor in migrant lowland rice paddies (*basakan*) and on palm oil plantations.[3] For many Pala'wan, relying on forests, fields, and rivers was increasingly insufficient. Like other highlanders, they contended with the vagaries of yield, markets, and outsider ideals of how life should be lived.

Nonstate Encounters

It soon became clear, however, that most Pala'wan households I encountered in Kamantian were entangled in another, far more pernicious social existence. One evening, as I sat in the entranceway of Panglima Kulibit's thatch house (*kubo*) overlooking the house yard (*legwas*), I noticed two young Pala'wan women, illuminated by new solar-powered lamps, weaving shaven and thinly stripped rattan and bamboo late into the evening. Kulibit explained that these women and other families were now making sizable volumes of the culturally significant *tingkep* baskets (and other customary handicrafts), once used only for storage, hunting, and rituals, as souvenirs for the island's growing tourism industry. In passing, Kulibit suggested that the basket weaving was first motivated by the parastatal Palawan Tropical Forest Protection Programme (PTFPP) and various other environmental nongovernmental organizations (NGOs). Teams from these organizations had visited Kamantian to advocate for tingkep production as an "alternative" to exploiting forests for swidden and NTFPs. Further south, in the

sitio of Marenshewan in Bataraza, the local CI chapter also believed that tapping income from this curious, ubiquitous customary basket (see fig. I.1) could
advance the MMPL's conservation objectives.

It was not altogether unusual for upland families to weave handicrafts to
support their livelihoods.[4] More notable was that the Pala'wan families who
wove tingkep also enjoyed singing Christian hymns while doing daily chores in
Kamantian. Many of their younger children frequently walked together, quasi–
single file, from one side of the community to the other. While many uplanders

FIGURE I.1. Weaving Tingkep at night by light, 2018. Photo source: Dressler.

adopted lowland religious rhyme and reason (Paredes 2006; Macdonald 1992a), in my two decades of research in Palawan I had never encountered this social configuration. I soon received an explanation.

I was approached—seemingly out of nowhere—by a clean-shaven young white man who enthusiastically greeted me: "Welcome to Kamantian and our church. We're here to help. What's your name?" Johnathon, apparently from rural Ohio, explained that he was with the Adventist Frontier Missions (AFM), associated with the Seventh-day Adventist (SDA) Church. With a health clinic, church, school, and helicopter at their disposal, their mission set out to reform, in their words, the "primitive" and "unreached" among the Pala'wan. I eventually learned that the AFM began their evangelical reforms with the community's youngest members. The line of children I had seen traversing the village was attending church that Saturday. The AFM invoked the wrath of God to deter the Pala'wan from believing in forest spirits, rituals, and myths—the very customs that informed their livelihoods and material culture, including the tingkep.

The Uplands: Nonstate Spaces of Reform

Across much of Southeast Asia, nonstate actors increasingly venture to seemingly remote uplands regions often forgone by state actors (Li 2007; Bryant 2008). They often arrive uninvited and work intensively to "improve" the character and well-being of the Indigenous poor by overcoming declining subsistence, customs, illiteracy, lack of hygiene, and the burning of forests. These are not the "come and go" NGOs of a decade earlier (Bryant 2001, 2002, 2008). Nimbler, more focused, and better funded than many state authorities, a greater number and diversity of nonstate actors have moved closer to Indigenous peoples and stayed put, governing them with zealous conviction in upland settings (van Schendel 2005; Scott 2009). These nonstate actors do not supplant the state per se; instead, state agencies often partner with or outsource responsibilities to NGOs to get closer to highlanders (Clarke 2006). Increasingly well-funded and organized nonstate actors fill the voids and deficiencies of state governance. Defined by clear program mandates, NGOs establish themselves with regular staff, infrastructure, incentives, and strictures that repurpose customary relations and livelihoods with technical solutions and moral agendas of purity and progress in the uplands (Li 2016).

Who or what beckons nongovernmental actors to these remote areas? Are they welcome? How do highlanders respond? And why are they, rather than the state, more present in such frontier settings? This book answers these questions by examining how and why nonstate actors aim to reform the lives of uplanders along virtuous, moral, and ecological lines, targeting customary objects, beliefs, and livelihoods to forge a social existence of Christian purity and ecological sustainability. Uplanders may solicit support from nonstate actors and embrace new ideas, identities, and practices for potential political and economic opportunities (Chua 2022). However, the sustained influences of such reforms on Indigenous livelihoods and social relations remain ambiguous and deeply consequential for a people who have independently and sustainably managed ancestral forests and biodiversity for centuries. Moving beyond binaries of acquiescence/resistance, I critically question how forest-reliant highland peoples adjust their lives and livelihoods in response to deepening, intersecting environmental and religious reforms over time.

Kamantian's "hinterland" status has long attracted various people and organizations who assumed state roles and responsibilities, including conservation and religious nonstate actors. Despite adopting different ideals and tactics, their longer-term goals have converged in practice, forming de facto government in the southern highlands of Palawan—a region where state environmental rule has had comparably less influence, even post-MMPL.[5] Almost all the nonstate entities traversing this landscape believed in "enhancing" Pala'wan life and livelihood by reforming the social and material things that mattered most to highland families, but were also deemed deficient. It was no coincidence that the small, ubiquitous tingkep basket and related livelihood activities emerged as symbols of reform—of disdain and desire—that anchored nonstate interventions over time.

Indigenous highlanders' customs, material culture, and livelihood practices have become increasingly immersed in nonstate actors' deepening social and ecological reforms of identity, belonging, and citizenship in southern Palawan. CI and other environmental NGOs have sought to remake and valorize the tingkep's customary character by encouraging Pala'wan to make more baskets (and other handicrafts) to sell in lowland tourism markets and, consequently, to become less reliant on swidden and reemerge as sedentary farmers and forest guardians. Conversely, the AFM has facilitated a deeper proselytization of Pala'wan by *unmaking* the ritual life, behaviors, and livelihoods associated with the tingkep and other material cultures, striving to realign them with a modern religious order. Their practices involved various relations, ideals, and technologies that were all aimed at disabling older ways of living to discipline and improve upon Indigenous bodies, beliefs, and livelihoods, in ways that could be considered biopolitical.

The Biopolitical Question

A rich history of anthropological scholarship (e.g., Conklin 1954; Condominas 1977; Olofson 1980; Perez and Bukluran 2018; Rosales 2021) and Indigenous-Filipino praxis (Guieb III 1999; Rosales 2020) reaffirms the ontological basis of

Indigenous peoples' rights to land and livelihood in the Philippines and elsewhere in Southeast Asia. A burgeoning literature has shown that Indigenous peoples have occupied, used, and managed forest landscapes relatively sustainably by drawing on varied nontimber forest products and agroecologically diverse swidden mosaics for centuries (Brookfield and Padoch 1994; Cramb et al. 2009; Dressler et al. 2017). Far from static and unchanging, the complex socioecological relations that constitute customary practices, livelihoods, and identities remain central to Indigenous peoples' ability to renegotiate and retain access to and use of ancestral lands and forests amid contested state and nonstate governance in the twenty-first century. Indeed, Indigenous peoples negotiate the perceived opportunities and concerns that stem from governance promises and persuasion *through* the legacies of the lands and forests they hope to leave for future generations. Most do not simply abandon forest, farm, and custom to pursue progress through such reforms. Yet the same characteristics that have ensured Indigenous peoples' resilience and biodiverse territories (e.g., Indigenous knowledge, social institutions, and forest management) have allowed nonstate and state actors to reify, discipline, and exploit these "attributes" as racializing imaginaries of the Other in the Philippines for national agendas and other global imperatives (Li 2000).

The legacies of Spanish and American colonial law criminalizing swidden livelihoods and vilifying Indigenous peoples' ways of life (see chaps. 3 and 4) persist in Philippine society and contemporary reform programs that extend well beyond state institutions and geographies of rule. This book explores how and why these biopolitical histories have become reimagined and enacted through translocal, nonstate actors and their interventions in southern Palawan and elsewhere in the country. Going beyond recent ethnographies of governmentality and agrarian change (West 2006; Li 2007; Scott 2009; Bierman and Anderson 2017), this book examines how and why both religious and conservation motives intersect in practice to influence social identities, livelihood, and difference through changing social relations and (self)governance. It considers how and why the reform programs of nonstate actors explicitly target sociomaterial objects and practices with the greatest significance for Indigenous peoples, particularly around social reproduction, beliefs, and livelihoods. It details how and why the social and material basis of Indigenous peoples' lives and livelihoods mediate the impacts and outcomes of nonstate reforms in uneven political and economic conditions. It describes who falls in line, who resists, who remains ambivalent, and who leverages nonstate interventions over time and space. In short, the book works toward a political ecology of reform by tracing the lineage, convergence, and responses to intensifying mixtures of religious and environmental biopolitics that manifest among different peoples and places in frontier areas. It aims to explain the causes and consequences of the parallel rise and convergence of religious and environmental nonstate reforms in spaces seemingly beyond the reach of the state (Wilkins 2021).

By bridging analysis of material culture, livelihood change, and governance (Tsing 2005; Saguin 2016; Currie et al. 2021), I consider how the multivalent character of customary objects and livelihoods shape and are shaped by the biopolitical reforms of nonstate actors. In this sense, customary objects and livelihood practices consist of dynamic social and material relations and are subject to contrasting cultural meanings and interpretations based on the social position, power, and authority of different actors (Shepherd and McWilliam 2011, 192). Foregrounding the character and influence of sociomaterial relations and practices offers deeper insights into how they mediate the interplay of life, livelihood, and projects of reform in the uplands of Palawan Island.

The book traces how and why missionaries and environmental NGOs came to the same upland region with the converging biopolitical objectives of reforming Pala'wan social existence. Highly divergent representational politics and ontological frames are at play here. For the Pala'wan, upland materialities (tingkep) and livelihood practices (swidden) are the basis of survival, reproduction, and ontological existence. For the nonstate actors, they are pagan material objects (fetishized) and destructive practices (vilified) in need of urgent social and ecological reform (O'Brien 2002). I tell the story of this social and ecological reform through the practices of diverse political actors and through the social lives of Pala'wan, their customary objects, livelihoods, and social relations. In particular, I consider how the unique tingkep basket and its place in Pala'wan life and livelihood mediates nonstate reform agendas at key conjunctures across south and central Palawan (see fig. I.2). The basket, its makers, and its many uses enmesh human and nonhuman worlds to inform personhood, group identity, and social practices (Miller 2006). The tingkep's sociomaterial character—human and nonhuman worlds comprising forest species, patterns, and deities—encompasses diverse worlds and varied encounters. The basket's social place among the Pala'wan cuts across the pragmatic utility of everyday uses (e.g., storing money, rice kernels, and small bones) to supernatural uses like holding stone amulets (*mutja*) and hosting the powerful female deity Linamen and other nonhuman forest beings. Its social and material character mediates life, livelihood, and project agendas, revealing the complex contours of Pala'wan ways of life and how they are incorporated into nonstate reforms.

The social reform practices these actors promote intersect with and inform the particularities of upland spaces and wider sociopolitical and economic processes involving state discourses, laws, and policies (Massey 1993). The book also

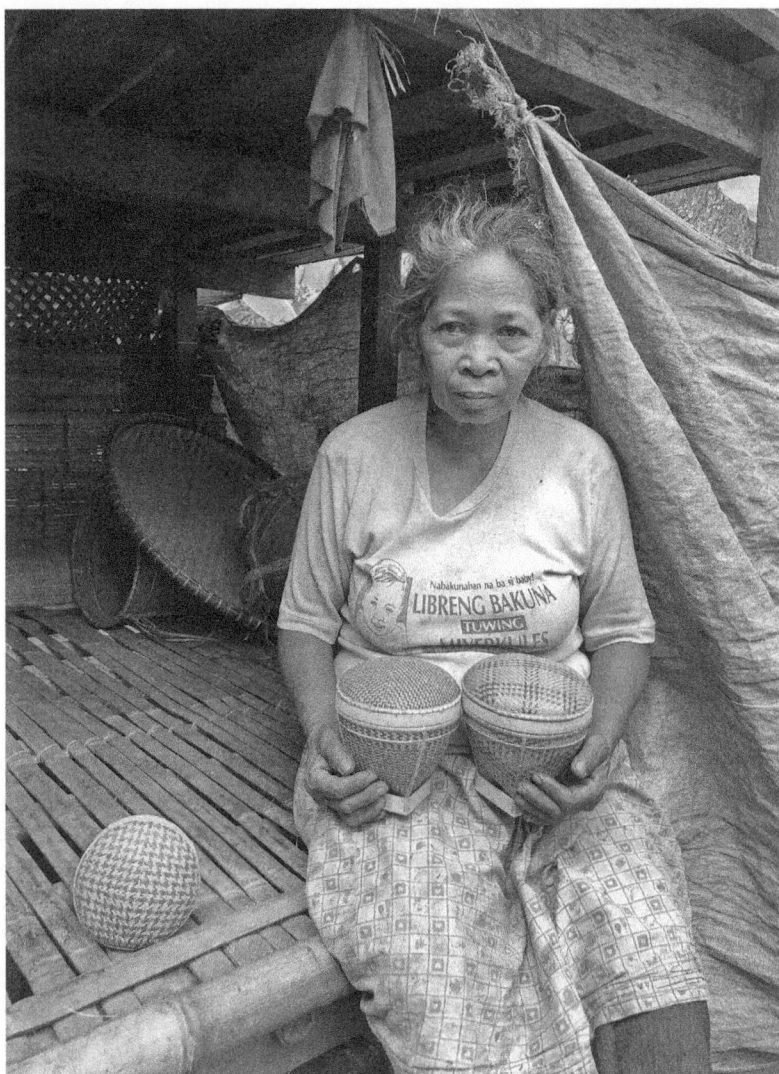

FIGURE I.2. Vital matter. The lidded Tingkep, freshly woven, 2018. Photo: Yayen.

considers how nonstate actors bypass and work with state institutions to exercise authority, legitimacy, and control to influence the conditions and possibilities of upland life over time. It considers the motivations, desires, and strategies that such nonstate actors bring to the uplands, the ways they are mediated through customary materials and beliefs, and the impacts and outcomes wrought upon Pala'wan lives and livelihoods.

For the Sake of Forests and Gods

The book's title, *For the Sake of Forests and Gods*, reflects the contrasting but mutually reinforcing governance objectives of forest conservation and evangelical missionizing. Both seek to unmake and remake uplanders' existence for the sake of ideological and moral ends. Both can be framed as religious, faith-based interventions driving social reform in the public and private realms of Pala'wan society. The subtitle, *Governing Life and Livelihoods*, captures the interconnectedness of such reforms that ostensibly "improve" uplanders' supposedly deficient and threatening ways of life through incremental social and material realignment. Ultimately, both are centered around the attempts—however partial or incomplete—to comprehensively reform Pala'wan existence. Nonstate conservation actors aim to remake Pala'wan life and livelihood by weaning them off swidden through alternative livelihood pursuits that overcome the supposedly primitive and criminal character of upland existence. Meanwhile, missionaries aim to expunge and replace all things customary, from rituals to hunting, with evangelical beliefs and practices of hygiene, morality, dietary purity, and monotheism.

Overall, this book hopes to represent the uplands as the Pala'wan might see such landscapes: as ancestral homelands filled with memories, lived experiences and aspirations. Perhaps more from my own (white, male) academic perspective than theirs, the book also aims to recast the Indigenous uplands as contested postcolonial spaces entangled in ideological expression, experimentation, and the political enrollment of broadly independent rural peoples already contending with uneven political economic realities (Eilenberg 2014; Lund 2021; Rasmussen and Lund 2018). Far from being marginal borderlands, certain upland spaces have become contested spaces where nonstate actors trial programs of reform among the "Indigenous poor" (Scott 2009; Korf and Raeymaekers 2013). Some uplanders engage with reform efforts, staying the course and hoping to seize opportunities to improve their lives. Some are indifferent to the lingering and limited project benefits. Others feel resignation and contempt. Others resist. Crucially, however, nonstate reforms often emerge without the invitation or consent of uplanders and are typically at odds with their sense of autonomy and survival.

This book therefore details Indigenous persistence, continuity, and change at these interstices of biopolitical reforms. Few Indigenous uplanders—relegated to the margins of state rule and neglect—have emerged as rights-bearing subjects with the same entitlements to property and protection as those citizens closest to the state (Lund 2021). They often embody Agamben's (1995) formulation of *Homo sacer*: those who are set apart from society and deprived of all rights. The inalienable, natural rights that nation-states supposedly afford citizens tend to diminish when subjects are framed as tribal Others beyond the remit of government. The many nonstate actors who ply the uplands—whether environmental NGOs or missionaries—now fill this governance vacuum with their own overzealous mandates of reform that aim to overcome what they perceive as suboptimal ways of being, believing, and subsisting. These mobile and motivated nonstate actors, powerfully informed by colonial histories, have emerged as the core "trustees" of Indigenous uplanders since the twentieth century (Li 2007). Working with or bypassing the state, they forge close relations and formal agreements with village leaders, closely monitor livelihoods and beliefs, and work toward disabling swidden lifeways by installing their own sense of modernity.

Situating the Cases

Much of the ethnographic work for this book took place during my repeat visits to the Palawan highlands over the last two decades. For this project, I spent four months in Kamantian and Marenshewan in 2013 and 2018 (see maps I.2 and I.3). Like most critical geographers who observe changing livelihoods and social practices in frontier regions, I sought out, documented, and reflected upon how social and environmental processes relate to, intersect with, and influence one another over time and space (Massey 1994, 2007). Using the analytic of "conjuncture" (see chapter 1; Li 2014), I aimed to understand how different "social, political, economic, and ideological contradictions" work to give the uplands "a specific and distinctive shape by com[ing] together, producing a crisis of some kind" (Hall and Massey 2012, 55). My method of following patterns in governance and local outcomes—from wider to narrow and intersecting contexts (Vayda 1983)—involved understanding what was, what is, and what may eventuate as processes and events converge in the uplands (Li 2014).

I focused specifically on the social and material relations of Pala'wan lives, nonstate programs, and the entangled political economic histories that shape everyday life in Kamantian and Marenshewan. As Massey (1994, 127) notes, "Places come to reflect the particular moments in such intersecting social relations, nets of which have over time been constructed, laid down, interacted with one another, decayed and renewed." As I show, social relations manifest in upland places and extend beyond them, "tying any particular locality into wider relations and processes in which other places are implicated too" (120).

I followed the tingkep's winding trade routes to understand how customary objects mediate biopolitical projects and change livelihoods and social relations across space. To "follow the thing" (Cook 2004), I reflected on how the basket's form, character, and value were reproduced through the meanings, identities, and uses of those producing, harvesting, owning, selling, and consuming it in rural and urban landscapes (Foster 2006). By following commodities through "diverse contexts and phases of circulation" (Foster 2006, 285), one can develop fuller stories of their existence (Cook 2004) or a situated "biography of things" (Kopytoff 1986). At moments of conjuncture, I describe how the basket and other customary items were transformed from socially embedded ritual objects and useful household items to partly disembedded commodities in the context of uneven production and exchange among makers, sellers, and agents of reform in urban settings.

After realizing the tingkep's importance to customary and reform agendas in Kamantian, I wanted to consider the basket's place and use in other villages across the Mount Mantalingahan Range. I followed the basket's trade routes across upland areas, to neighboring AFM sites, through the lowlands, and, ultimately, to the capital city of Puerto Princesa, where Pala'wan hawked the baskets, and souvenir shops hoarded, stored, and sold them in large quantities. I also visited Pala'wan in three different villages across the Victoria-Anepahan Mountain Range farther north (in Quezon Municipality). Most of the people there knew little about crafting the tingkep, and none knew about the myth and rituals surrounding the basket. Instead, they specialized in making large quantities of wooden masks for tourism souvenir shops in Puerto Princesa. Other villagers wove the *tabig*—the larger lidless rattan basket used for dry-storing root crops, rice, clothes, and other items—or the *sukatan* for carrying heavier loads of crops or game harvested from swidden fallows and forests (see fig. I.3).[6] Most Pala'wan who wove and used the tingkep lived farther south along the eastern ridges and slopes of the Mount Mantalingahan Range near Marenshewan and Kamantian. In each area, I visited several remote household clusters, such as Lap Lap (see map I.2). The remainder of my time was spent in the Victoria-Anepahan Mountain Range, the Quezon region, and Puerto Princesa City.

Kamantian

Kamantian is located between eight hundred and one thousand meters above sea level, at the end of a three-hour-long (only ninety minutes for Pala'wan) hike along a narrow forest trail. The sitio itself lies within the Barangay of Salogon and the Tamlang watershed (of Brooke's Point municipality), a bowl-like depression flanked by steep mountains with eroding frontal slopes (see fig. I.4). The moderate to flatter sections of the watershed host approximately thirty household clusters with trails that radiate outward to swidden mosaics—cleared and burned fields or plots at various stages of fallow regrowth. The Pala'wan of Kamantian are a mixture of horticulturalists, hunters, and gatherers who use

FIGURE I.3.   The Sukatan basket, 2018. Photo: Yayen.

products from their swiddens for subsistence, cash income, or both. They intensively harvest various NTFPs, such as wild pig (*biek talun*), rattan (e.g., *uwey*,
nawi), almaciga resin (*bagtik*), and honey (*deges*) in swidden fallows and the
broader forest landscape. These products may be sold to lowland buyers for cash
or credit, or used to make items like the tingkep; other customary objects (e.g.,
musical instruments like the two-string lute, *kudlung, kudyapi*); household items
(e.g., hunting blowpipes, *sapukan*); house walling (*sawali*); and roofing (*pawid*,
nipa palm). Many Pala'wan also pursue intermittent wage labor opportunities
in migrant paddy fields, copra fields, oil palm plantations, and mixed construction to supplement their incomes. The main subgroup of Pala'wan informing
this study comprises a cluster of approximately twenty households, part of the
Mekagwaq-Tamlang area. Other households are located farther into the interior of the watershed, some distance away from the Adventist camp (Macdonald
2007; see map I.2).

Moving downslope and westward from central Kamantian, a lower section
of Pala'wan homes fringes the well-maintained, curated AFM area devoid of
swidden fields. The AFM facility is complete with bunkhouses for young women
disciples, a church, a schoolhouse, a clinic for sick Pala'wan, and a laboratory, a

FIGURE I.4. Kamantian and the Tamlang Valley, 2013. Photo: Dressler.

biopolitical operational space typically the remit of state agencies (see fig. I.5). AFM nurses, a pilot, volunteers, and the pastor and his wife live in well-maintained houses constructed of costly sawn hardwood timber and cement foundations, flooring (*anibog, Oncosperma filamentosum Blume*), walling (*sawali*), and tin roofing. An aircraft runway and helicopter landing area sit further downslope. When operable, the helicopter takes the AFM nurse, doctor, or pastor deep into the interior for medical and proselytizing missions (see chaps. 5 and 6). In other cases, desperately sick Pala'wan are airlifted to the Kamantian clinic or better-equipped hospitals at Brooke's Point or in Puerto Princesa City. The helicopter also brings medical provisions and goods to keep the clinic and mission running.

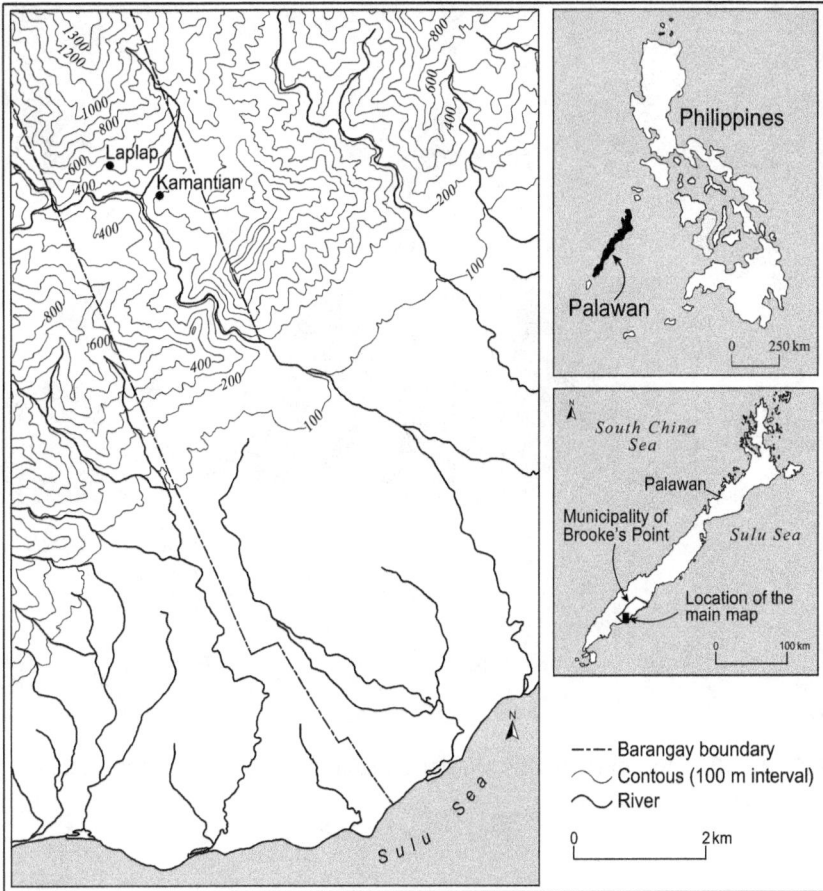

MAP I.2. Kamantian, Brookes Point, Palawan. Cartography © Chandra Jayasuriya, 2022.

During my stay, Pala'wan laborers were constructing sturdy mission-financed homes (cut timber, concretized bases, and tin roofs) for Pala'wan disciples near the runway (see fig. I.6).

Four Pala'wan families who had distanced themselves from the mission lived farther upslope in the small sitio of Lap Lap. My visit to Lap Lap was brief but immersive. Upon arrival, I pulled my back and was treated by a Pala'wan *beljan* (ritual specialist, healer) with incantations and the sweet fragrance of *ruruku* (basil, *Ocimum sanctum*). He identified the cause of my pain and withdrew a stone that had apparently lodged in my back. After this, I felt well enough to continue my research.

FIGURE I.5. Adventist Frontier Missions church and school, Kamantian, 2018. Photo: Dressler.

Marenshewan

Located about eight hundred meters above sea level, the sitio of Marenshewan, Barangay Bono Bono (of Bataraza municipality), consists of several housing clusters belonging to three main families, their extended kin, and others unrelated to them (see map I.3). At the time of my research, the ten households were located across Marenshewan proper, downslope, and farther into the interior of the Mount Mantalingahan Range. As in Kamantian, the Pala'wan here are dedicated swidden horticulturalists—rice and root crops are their subsistence mainstay—who also hunt, gather, collect NTFPs, fish, and work as day laborers in the lowlands.

The Marenshewan panglima, Franco Bilog, was the local broker for CI. He often convened community meetings for rituals, arbitration (*bisara*), and NGO-related discussions in the newly built, multipurpose tribal hall (see building with tin roof in fig. I.7). CI built this structure as an alternative learning center and environmental education facility for both young and old Pala'wan. In the

FIGURE I.6. Cut timber and concrete foundation houses for AFM disciples, 2018. Photo: Dressler.

shadow of the CI building, Pala'wan continued to cautiously cultivate swidden in young secondary growth forests (see fig. I.8) and "underbrush" (as opposed to "hot burn") fields that hosted fruit trees and hardwoods—part of the NGO's tree planting and swidden eradication initiative. Bilog and other Pala'wan from

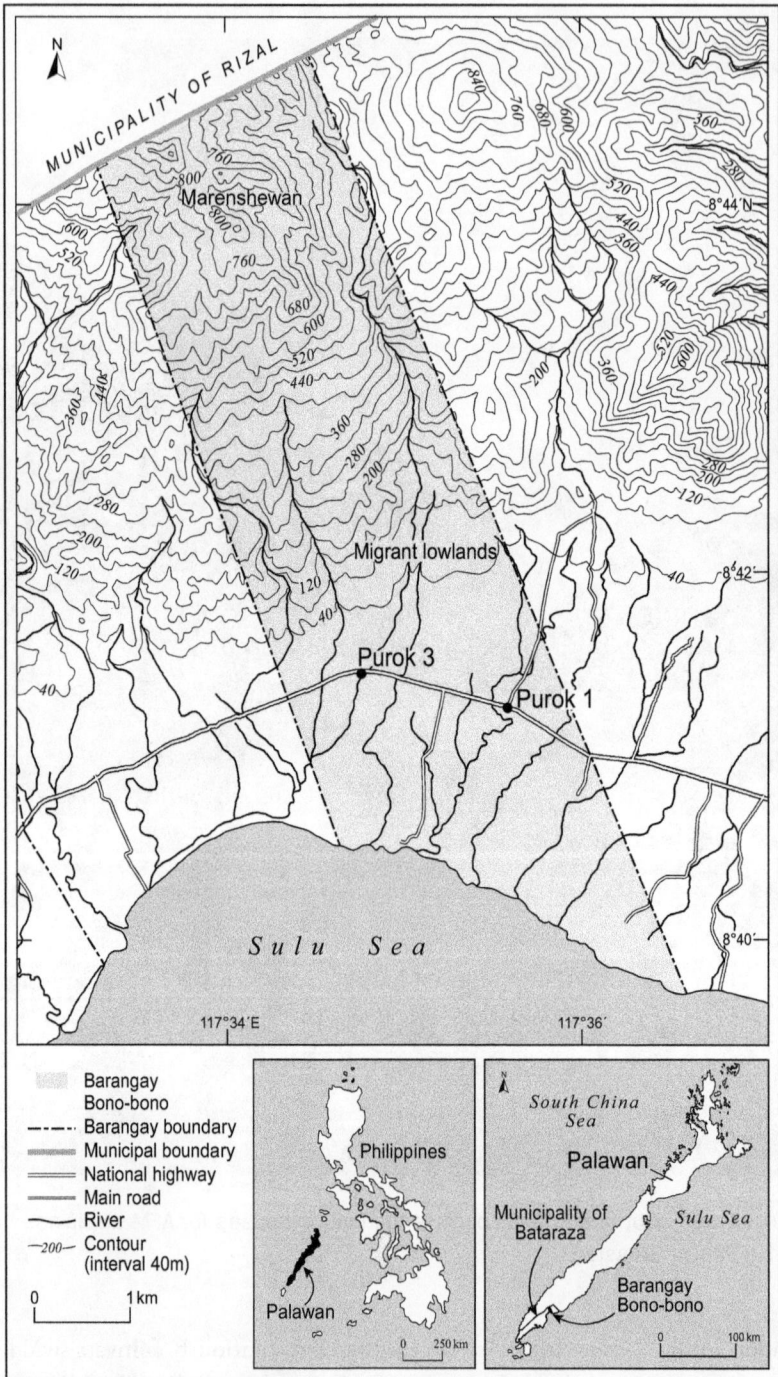

MAP I.3. Marenshewan, Bataraza, Palawan. Cartography © Chandra Jayasuriya, 2022.

FIGURE I.7. The Marenshewan uplands, "Tribal Hall," and other buildings, 2013. Photo: Dressler.

FIGURE I.8. A cleared uma (swidden field) in the Marenshewan uplands, 2013. Photo: Dressler.

the area frequently appeared in glossy CI media material and reports on planting trees in swiddens, monitoring reforested areas and clearings, and weaving tingkep baskets to supposedly offset any lost income from livelihood substitutions.

On Method

In both Kamantian and Marenshewan, I drew upon a range of methods to understand Pala'wan livelihood characteristics, NGO associations and their interventions, as well as the crafting, meanings, and uses of the tingkep. I spent most of my time with Pala'wan family members involved in nonstate programs, known experts in swidden farming, and those knowledgeable about the tingkep's roles in myth, ritual, and livelihood. Discussions mainly centered on how tingkep were made and from what materials, the meaning of weave patterns, size, and color, as well as how these aspects influenced the basket's use across human-nonhuman relations. Over time, I became more interested in how the use and presence of the tingkep were situated at the intersection of contrasting livelihood priorities, beliefs, and reform efforts in both upland and lowland areas.

In Kamantian, I conducted semistructured interviews and oral histories on the place of tingkep in the human and nonhuman worlds of the Pala'wan. The oral histories narrated how its social and material character centered on myth, ritual, religion, supernatural worlds, and forest landscapes. In Marenshewan, I completed similar interviews about the tingkep and Pala'wan livelihoods. In both areas, I spoke at length with elderly female craft weavers, some of whom have since passed away. These hour-long discussions covered the social and material dimensions of the crafting process, the use of different baskets (types and sizes) in different social settings, and the use of tingkep in ritual—particularly the Tingkep 't Kundu ritual, which was still practiced in villages near Kamantian.

Between visits, I developed a simple livelihood survey to document the basket's position in livelihood activities in the uplands and lowlands of the region. I surveyed ten (men and women) heads of households in Marenshewan and twenty households in Kamantian (a near-complete sample) to understand the main Pala'wan livelihood activities, the volume of baskets made and sold, and the income generated relative to other pursuits in the 2012 calendar year. Among the various livelihood pursuits, I surveyed swidden yields (across fields with varieties of upland rice, root crop, legumes, and tree crops in fallow) for subsistence and sales, daily wage labor (e.g., on lowland paddy fields and SDA plots), the harvest of NTFPs, and the crafting of baskets and other handicrafts for home use and sale (as souvenirs in lowland tourism markets).[7]

I also sought to understand how two nonstate actors (the CI and AFM) and one parastate actor (the PTFPP, which uplanders tended to perceive as an NGO operator) imposed different controls, strictures, and incentives upon the Pala'wan with overlapping logics and outcomes. My discussions with former PTFPP and CI staff and project beneficiaries were revealing. Current and former staff members spoke candidly about their efforts to establish social contracts, so-called community conservation agreements, with Pala'wan for long-term forest conservation. These agreements enrolled Pala'wan in alternative livelihood strategies, environmental education, and training designed to draw them away from existing swiddens that, ironically, pushed them to clear older forests and overuse NTFPs in tingkep production.

It was nearly impossible to speak candidly with AFM members, particularly the American missionaries in Kamantian. Many of them scrutinized my background, motives, and intentions before ultimately refusing to be interviewed at all. During my final visit, the AFM pastor's wife not only questioned me about my discipline but also wanted to know more about changes in Pala'wan mythos, lore, and livelihood. I suggested that they surely knew more than I about the Indigenous peoples they were attempting to reform. Such awkward encounters were rather common.

As Bonsen et al. (1990, 150) note in *The Ambiguity of Rapprochement*, meetings between ethnographers and missionaries are almost always awkward and tense, irrespective of whether the parties are foreign or local. Both sides consider themselves to be experts and "representatives of a foreign culture" to which they do not belong, and each, in their own way, makes an ideological case for which social attributes—or ways of life—should be supported, privileged, or removed. Similarly, van Beek (1990) explains that these "ethnographic moments" are not all lost in translation: ethnographers and missionaries read each other's cultural scripts, partly to progress their own mission. Such semantic moments reflect a political conjuncture or, as van Beek calls it, the emergence of cultures in-between.[8]

Most of my interactions with the AFM entailed casual discussions with the Pala'wan faithful. They explained the relative benefits of the faith and mission, while other Pala'wan farther upland frequently spoke of "backsliding" from the faith for mundane, everyday reasons. Others failed to appreciate the AFM's presence. I also collected information about AFM practices from publicly available online literature, detailed blogs, and videos of the AFM in Kamantian (e.g., as presented by mission staff on the AFM website, YouTube videos of Kamantian, pamphlets, online Adventist blogs of interventions in Kamantian).[9]

After leaving the uplands, I spent several weeks in the provincial capital of Puerto Princesa City to follow the tingkep's movement through social networks of production and exchange. I visited major urban and periurban markets, such as the New Public Market, to interview Pala'wan traders who sold crafts on weekly runs (e.g., *walis tambo*, or brooms, and tingkep). The traders traveled from upland areas and often remained in the city for several days to sell their goods.

I also interviewed twelve souvenir shop owners who bought, sold, and collected Pala'wan crafts, including the tingkep. Interviews with store owners and managers were relatively straightforward, since most souvenir shops are located along Rizal Avenue, the city's main tourist drag. I distributed a questionnaire to record the tingkep's buying and selling prices, the quantity stored, the baskets' origins, the means by which they were obtained, and the shop owners' perceptions of the baskets, as well as of Pala'wan weavers and Indigenous peoples, generally. I also visited their storage areas—stuffy backrooms that held mountains of old Indigenous crafts and customary items covered in dust and mold and generally in poor condition. Moving between the uplands, lowlands, and the city allowed me to trace the tingkep's movement between key actors and across different spaces to better understand its changing value(s) at different points of production, exchange and consumption. Consent was given and recorded for all methods of data collection in all study locations.

Organization of the Book

This book unfolds chronologically, often pivoting between an analysis of regional and local areas. Chapter 1 outlines the theoretical and conceptual framing that gives this ethnography traction across the diverse empirical themes explored in subsequent chapters. Chapter 2 introduces Pala'wan livelihoods and worldviews, including the tingkep's vitality across social relations, livelihoods, and forest worlds. I detail the basket's mythic origins, sociomaterial character, and ritual substance.

Chapter 3 critically reviews key moments of Spanish colonial rule, especially the Catholic Church, state, and military's civilizing missions to pacify, Christianize, and sedentarize uplanders. Such reform programs governed access to and use of forests along racialized socioecological hierarchies. Indigenous peoples were constructed as dark-skinned, primitive, upland tribals who led criminal, "slash and burn" lifestyles. Yet the chapter also notes the limited reach of Spanish colonial administrators and the Catholic Church in reforming Pala'wan uplanders, who were already involved in complex tribute and trade with Tausug lowlanders in the Muslim south of the island.

After defeating the Spaniards in 1898, American colonists reworked this ideological and institutional legacy into a more comprehensive and sophisticated system to reform Indigenous peoples' lives and livelihoods through efforts to codify and administer lands, forests, and recalcitrant highlanders. Chapter 4 describes how customary beliefs, poor hygiene, and degrading swidden were politically constructed as inferior, primitive, heathen characteristics necessitating deeper reforms. Despite the American colonial administration's relative effectiveness in partly suppressing swidden and producing hygienic "Christian natives," the colonial government later used Baptist missionaries to reach the "unreachable." Lamenting the Catholic Church's half-hearted proselytization, the SDA and other Baptist missions soon worked to discipline Indigenous uplanders into adopting evangelical beliefs and lowland ways of life.

After World War II, the forestry departments in the Philippines doubled down on community-oriented efforts to eradicate swidden farming, sedentarize uplanders, and render civilized subjects. By the 1980s, nonstate actors had established themselves across the country's frontier areas. Environmental NGOs and evangelical missions increasingly worked in the same landscapes, both bypassing the state in governing Indigenous uplanders on Palawan. While these nongovernmental entities governed highlanders by living among them, they continued to uphold the same ethnoreligious, racialized practices of the colonial era. These biopolitical strategies are examined in depth in the extended and integrated case studies of the next three chapters.

Chapter 5 examines how these historical conjunctures coproduced "marginal" uplands and "primitive tribals" to constitute spaces of experimentation and objects of reform for nonstate actors in the contemporary period. I then describe how today the political economy of nonstate actors on Palawan involves targeting, reforming, and optimizing Indigenous uplanders through key social and material markers of custom and indigeneity, including the tingkep basket and its associated practices. Chapters 6 and 7 ethnographically examine how environmental NGOs and missionaries design and perform in their respective biopolitical theaters (e.g., administration, technology, infrastructure, knowledge, incentives, and strictures), and how Pala'wan respond to intersecting designs that aim to produce environmental and religious subjects. In theory, raising Pala'wan consciousness of their primitive and degrading conditions would make them perpetual adherents to nonstate ideals and practices. In practice, efforts to raise the Pala'wan consciousness concerning their "condition" were defined by tensions and contradictions.

In both chapters, I describe the implications of nonstate actors imposing their ideals and practices upon Pala'wan lives and livelihoods in the context of uneven political economies. I examine the tingkep's enrollment in and influence on these interventions, wherein its use, reification, and optimization for craft production schemes were meant to enhance the sustainability of Pala'wan livelihoods. In chapter 6, I examine how environmental NGOs such as CI indigenized tingkep "branding" to generate income from basket sales to subsidize the loss of swidden. In chapter 7, I detail how AFM evangelists, "deep reformers," sought to transform the personal, intimate spaces of Pala'wan lives and customary materials—including the tingkep and associated social practices—by "purging the devil" and bringing them "closer to God." These intersecting reforms were partial and contested, oftentimes refracted through acts of resistance or indifference from Pala'wan households. Each chapter shows how critical reflections drove a wedge between households, the Adventists, and environmental NGOs. Overall, the chapters show how and why these nonstate entities ended up adopting de facto state roles and practices, reaching into the private and public spaces of the Pala'wan in Kamantian and Marenshewan to undermine their sovereignty and self-determination.

I conclude by reflecting on why biopolitical reform projects often thrive but with varying consequences in upland spaces. Many Indigenous peoples consider the marginal uplands to be their ancestral homelands. Here, they negotiate and take advantage of the liminal spaces that exist between state and nonstate projects of rule (Das and Poole 2004; Korf and Raeymaekers 2013; Scott 2009; Dressler and Guieb 2011). However, partly because state bureaucracies reproduced essentialized and reified notions of Indigenous peoples, nonstate actors feel politically and morally obligated to descend upon hinterland areas to offer up reforms in the relative absence of government, or, sometimes, because of its presumed failings. Nonstate actors gaze upland with a moral authority to represent the marginal while reinforcing and reforming naturalized cultural differences forged from centuries of colonial rule (Anthias 1998; Igoe 2005).

These short and longer-term reform projects are based on a long-standing politics of difference. Program motives are tied to identity, culture, livelihood, body, and soul in ways that reproduce interventions beset with contradictions, contestations, and violence against uplanders. Though uplanders may not always fully understand the motivations and implications of those outsiders attempting to govern them, when they do they may or may not agree with the motives and outcomes. Programs of reform usually intersect in upland areas in ways that ultimately undermine the sovereignty of Indigenous peoples living there. I hope that those reading this book—whether state official, NGO practitioner or student, in the Philippines or elsewhere—will critically reflect upon the political contradictions and consequences that nonstate programs have had on upland peoples who wish to negotiate modernity on their own terms and in line with their own aspirations.

1

BIOPOLITICS, MATERIALITIES, AND
THE POLITICS OF DIFFERENCE

> Pastoral power initially manifests itself in its zeal, devotion and
> endless application. . . . The shepherd is someone who keeps watch.
>
> —Michel Foucault (1978)

The uplands of Southeast Asia have long been framed as unruly, marginal, and backward spaces that require disciplining through modern ways and means, whether through the watch of the pastor or the project broker (Van Schendel 2002; Van Schendel and De Maaker 2014). Framed through colonial frontier imaginaries of marginality and surplus (Barney 2009; Fabinyi et al. 2019), the uplands have reemerged as contested political arenas in which nonstate actors aim to reform "vulnerable" forest dwellers in accordance with contrasting social and ecological ideals. This chapter outlines the concepts that underpin the book's narrative of Indigenous peoples' struggles, contestations, and alignment toward nonstate practices of reform within and beyond the uplands of Palawan Island. It shows how, for the Pala'wan, the southern highlands remain ancestral territories that are replete with cultural memories, livelihoods, and struggles that intensify as nonstate interventions arise (Ocampo 1996).

I first describe the book's overarching analytics of conjuncture and biopolitics and then the subthemes of social reproduction, materiality, and the politics of difference. Together, these concepts help us to understand how nonstate practices intersect with and influence the lives and livelihoods of the Pala'wan and other Indigenous peoples in the region. They shed critical light on how biopolitical practices reproduce and perpetuate—at the nexus of conservation and evangelism—notions of modernity that aim to reform the social and material basis of Indigenous existence in terms of nonstate ideologies and social practices. I develop these subthemes in subsequent chapters to investigate how and why the social and material relations of the Pala'wan mediate nonstate programs and practices in the uplands.

Political Conjunctures

Across the uplands of Southeast Asia, a diverse range of nonstate actors and their interventions have expanded in reach and intensity, accelerating how social and political moments converge, transform, and produce new possibilities and constraints over time (Clarke 2014). Further developed by Stuart Hall, the conjuncture analytic involves tracing and examining how a "number of forces and contradictions, which are at work in different key practices and sites in a social formation, come together or conjoin in the same moment and political space" (Hall 2011, 705). Conjunctural dynamics refer to the tensions, contradictions, and opportunities generated by the intersection of different actors' interests, ideologies, and practices over time and space. Building on Hall's framing, I use conjunctural analysis to examine how historical and contemporary nonstate (and state) reforms converge as a "multiplicity of forces and accumulated antagonisms" (Clarke 2014, 115) that influence how Pala'wan make a living, engage with material culture, and identify (or not) with those trying to govern them.

"Biopolitical conjunctures" involve nonstate and state ideas, practices, and social relations that "articulate together at a particular locus" (Massey 1994, 154) to influence how Indigenous peoples ought to live in changing forest landscapes. Although these biopolitical ideals and practices might be fragmented and reframed (Cepek 2011), they undoubtedly affect how Indigenous peoples understand themselves, live, and make a living. Uplanders themselves often reflect on the meaning and impact of the programs and projects that become part of everyday life. As they negotiate these interventions, Pala'wan may begin to question what it is they have agreed to and whether project promises will be fulfilled. Through a Foucauldian lens, I interrogate how and why nonstate actors and their reform agendas have remained so determined to improve and remove what the Pala'wan already know and have been doing so well.

Biopolitics and Pastoralism in the Uplands

The once apparent distinctions between governance interventions designed to enhance human well-being and those aimed at improving sustainability or biological health have progressively blurred (Cavanagh 2018, 410). In recent decades, the various state or nonstate actors pursuing such interventions have drawn on their own logic, power, and authority to enable reforms beyond narrowly targeted disciplinary practices (Fletcher 2017).

Foucault (1978) famously described how biopolitics (or biopower) entails strategies and interventions that center on enhancing life itself (Lemke 2011, 33). Biopolitical practices manifest as a blend of disciplinary and regulatory interventions, spanning the discursive, subjective, and biophysical realms of individuals and populations (Foucault 1978, 139). Disciplinary interventions centered on optimizing the physical capabilities of individual bodies, leading to a parallel increase in social potential through more efficient economic systems of control (139). Achieving this objective involves overcoming and enhancing the emotional, physical, and material conditions that constrain individual existence (139). Concurrently, a range of regulatory controls operate through (or independent of) a set of disciplinary practices, creating new institutions and behavioral norms that influence how people understand and manage the biological basis of their own population (Foucault 1978, 139). Early literature in political ecology and allied disciplines (Agrawal 1999; Dressler 2014) often presumed that as Indigenous peoples appeared to align with the values and beliefs of technologies of rule, they would eventually progress toward self-regulation and collective governance in accordance with the wishes of nonstate actors.

In a deeper biopolitical sense, nonstate actors aim to govern highlanders' beliefs, behaviors, and actions through the "conduct of conduct": activities "aiming to shape, guide or affect the conduct of some person or persons" (Gordon 1991, 2). Such governmental practices employ diffuse forms of biopower to affect peoples' beliefs, understandings, aspirations, and actions through consensual means (Nelson and Wright 1995, 9–10). They include a complex suite of technologies, beliefs, strictures, incentives, and practices that aim to govern or "structure the possible field of action of others" (Foucault 1983, 221). Nonstate actors govern patiently and persistently, to avoid displaying overt strength and superiority. Without openly maligning free will, they extend a type of pastoral care through the "zeal, devotion, and endless application" of ideals and strictures on Indigenous bodies and thoughts (Foucault 2007, 127). Such endeavors reflect a duty of care—or pastoral power—applied to seemingly helpless people in vulnerable conditions. Mission and NGO staff emerge as benevolent pastors, working hard to ensure that their local sheep (i.e., project beneficiaries) are elevated to a social and biophysical level that overcomes the suffering they endure from their own seemingly primitive, backward ways (2007).

Conservation organizations and missionaries have emerged as the biopolitical experts of their time, influencing lives and livelihoods through individual and group reforms in remote places where the state struggles to govern. They regularly ply the highlands to locate suitably degraded forests and vulnerable peoples to establish complex infrastructure and programs to modify and improve upon their conditions (Lemke 2011, 33). Perhaps better than any state agency, nonstate actors have developed the uncanny ability to embed themselves within Indigenous communities to closely regulate and discipline the personal and intimate affairs of the social body—who I am and what I must become—and the wider political economy (Berlant 1998).

In the forests of southern Palawan, we see how nonstate actors work through the intimate and personal spaces of upland peoples' homes and village settings. They seek to establish closer relations with the Pala'wan in the hope of reproducing "aspiration[s] for a narrative about something shared" and agreed upon (Berlant 1998, 281). Such practices are carefully designed to instill a sense of private (inward) familiarity and comfort, along with public (outward) consent and sanctioning, to reinforce wants and desires that align with modern presuppositions. Nonstate biopolitical influence transcends the private, personal realm and the outward, public realm of upland households, communities, and social institutions. Caught in the trappings of post-Enlightenment modernity, environmental NGOs and missionaries attempt to forge modern subjects through a preconceived notion of temporality in which individuals and communities are improved by being led away from a "less civilized" past toward a less "burdensome future" (Hubinger 1997; Lee Dawdy 2010). Such a temporal ideology valorizes progress through mundane acts of substitution alongside profound acts of erasure (Lee Dawdy 2010, 762).

However, biopolitical practices are far from totalizing. Few Indigenous uplanders become fully fledged subjects, disciplined by others and themselves. Indeed, scholars have cautioned against using Foucauldian governmentality to describe overtly coherent and systematic "mentalities of rule" in conservation and development practices (O'Malley, Weir, and Shearing 1997, 501; Cepek 2011). They argue that by using Foucauldian concepts in a totalizing manner, one risks assuming a priori that processes of interpellation, self-regulation, and aligned behaviors coemerge. The impacts and outcomes of such practices may be overstated, sacrificing more open understandings of how Indigenous social practices refract or (re)appropriate such subjectification for strategic ends (O'Malley, Weir, and Shearing 1997; Cepek 2011; Chua 2022).[1]

Although I agree with these sentiments, I am equally skeptical of the suggestion that local subversive practices such as "critical consciousness" (Cepek 2011), everyday acts of resistance (Scott 1986), and reappropriating religion as a "political project" (Chua 2022) easily counter the friction of biopolitical interventions and the constraints they impose on human agency. Leaning too heavily on how social agency (and subaltern ontologies) might complicate the biopolitics of governance risks neglecting the influence of such reforms. The character and persistence of biopolitical interventions (e.g., actors, form of controls, incentives, and types of disciplining) and how they converge with other political economic processes and upland life determine the degree to which such agendas influence social practices, livelihoods, and ecologies over time. Biopolitical reforms are often neglected, avoided, or simply fail. At specific conjunctures, however, such performative powers can be exerted with and through other pressures and opportunities, amplifying affective outcomes over time. Even if partial and refracted, some biopolitical practices nevertheless stick (Dressler 2014).[2]

While Foucauldian biopolitics offers critical insights into how and why nonstate reforms affect the worlds of uplanders, the concept alone fails to adequately account for the variety of social and material responses to interventions and livelihood changes in upland settings. How, for instance, do varied governmental efforts intersect with and influence the complexity of everyday life and livelihood in upland settings? A deeper understanding of how Pala'wan households respond to reform projects amid ongoing agrarian changes requires additional insights into how nonstate interventions work through core aspects of Pala'wan social existence: ongoing livelihood changes, sociomaterial relations, and social reproduction. The following sections consider how unrelenting nongovernmental desires to modify and improve the conditions of highlanders articulate—or collide—with Indigenous material cultures, livelihoods, and beliefs over time and space.

Social Reproduction

Nonstate actors have become specialists in interpreting, managing, and steering the ways in which Indigenous peoples reproduce themselves within and through the social, biophysical, and ecological conditions of upland settings (Li 2007; McDougall 2020; McElwee 2016; West 2006). They focus incessantly on the various social relations, biological systems, and livelihood practices that coconstitute "social reproduction" (Katz 2004, 19) across gender and generation (Katz 2004; Bhattacharya 2017)—that is, "the labour and social processes and relationships that support production, exchange, and the maintenance of individuals, households, and communities" (Fernandez 2018, 145). Social reproduction also includes biological reproduction across generations of laborers, the reproduction of labor power in and beyond the domestic sphere, and the reproduction of labor as a social class defined by ethnicity, culture, and livelihood—all of which fall into nonstate actors' governance ambit.

Both environmental NGOs and missionaries typically begin their reform programs by identifying (neglecting or rendering invisible) how and why Indigenous peoples reproduce themselves (Miller 2005; Eder 2006; Kofman 2014). They almost always draw on long-standing assumptions about what attributes of upland living (e.g., social relations, livelihood labor) need to be reformed to justify interventions, objectives, and legitimacy. The targeting of certain attributes may involve categorizing, monitoring, criminalizing, depleting, or valorizing to facilitate deeper alignment and reforms. Conservation-oriented NGOs will use temporary project interventions with "a fixed goal and budget [and] technical matrix" to gauge "reproduction" within "degrading" livelihood practices (Li 2019, 1). Missionaries may aim to "live with" and understand reproduction for deeper reform of emotions, beliefs, biological processes, and livelihoods in the territories they occupy. For both groups, then, understanding and reworking the substance of Indigenous social reproduction is central to the success of "conversion" processes (Beatty 2012).

An analysis of how such conversion transforms processes of reproduction can reveal not only "a [deeply emotional] revision of boundaries, an unsettling switching of sentiments" (Beatty 2010, 298) but also the fact that household reproductive labor has long been influenced by uneven market production, ideologies, and institutions of reform (Bhattacharya 2017, 5–6).[3] These interpretations of "social reproduction"—despite partly diverging from Foucauldian thinking—involve an analysis of continuity and change among uplanders as their social realities entangle with overlapping reforms across scale. As much as Pala'wan social reproduction is part of ancestral lands and forests, it has also long been part of, and influenced by, sociopolitical and economic processes that cut across uplands, lowlands, and metropolises (Douglass 2006, 423; Peluso and Purwanto 2018).

Fixed Households and Project Beneficiaries

Nonstate actors who ply the uplands of Southeast Asia often deliberately align their programs with the most outwardly recognizable forms of social organization. One of these is the ostensible "household unit." In the Philippines, nonstate actors frequently target and enroll households in projects based on preconceived assumptions of family, kinship, coresidence, and shared labor that reflect the archetypal nuclear family. Environmental NGOs, in particular, have a long history of constructing and enrolling notional households—or heads of households—as "participants" and "beneficiaries" in complex conservation agreements. With households in the viewfinder, they focus on how best to use family labor for interventions and how reforming labor relations can steer families away from seemingly destructive livelihood practices such as swidden cultivation and hunting game (Greenleaf 2021; Neimark et al. 2020). For many, the most expeditious and enduring way to align individual, family, and household labor with project aims and objectives is by enrolling Indigenous children and parents in age- and gender-specific reform practices. Tree planting initiatives now often work through childhood environmental education and religious schooling in the remotest of areas (see chaps. 5–6). The assumption holds that family labor and householding are synonymous (Wilk 1989), that they work in unison, that they will support interventions, and that such "cooperation" exists across a community.

In the world of nongovernmental program design, the putative household persists as a nearly universal social unit and the most common form of famil-ial social organization and reproduction (Netting 1993; Wilk 1989, 1991; Eder 1999; Wolf 1991). For many NGOs, the durability and utility of the so-called household unit persists because it appears to retain the quasi-corporate char-acter of fixed coresidence, decision-making, moral expectations, consumption, distribution, exchange of things, and other types of reproduction (Netting 1993; Eder 1999).[4] Drawing on such universal categories, nonstate actors are further influenced by state law and religion to often mistakenly choose to enroll "de jure" heads of households (i.e., men) instead of the "de facto" heads (i.e., women) in programs and projects (Wolf 1990). Certain NGOs thus retain and insert power-ful Eurocentric assumptions about familial organization and production into their project design and execution.[5] These biopolitical assumptions persist and often misalign with upland life on Palawan Island.

State and nonstate assumptions about household organization and repro-duction fail to map neatly onto Indigenous familial organization, livelihoods, and social relations. As Eder (1999) observes, paying attention to Indigenous household characteristics like age, gender, authority, and social practice reveals a complex picture of social organization, reproduction, and coresidence. Many correspond neither with fixed household formations and functions nor with dominant imaginaries of collective, cohesive, and homogenous Indigenous communities (Dressler et al., 2010). Many highlanders like the Pala'wan disrupt static and narrow interpretations of familial reproduction, livelihood, and vil-lage residence (Macdonald 2007, 2011, 2012). Indeed, Pala'wan social relations, organization, and residence patterns are broadly acephalous or nonstructural (Revel et al. 1998). Such relations are defined by a relative absence of leader-ship or hierarchy—few Pala'wan have the authority to give orders (Revel et al. 1998; Macdonald 2007)—alongside strong principles of sharing and equity. Residence patterns also vary over time; any number of family members and individuals may live together, arrive, settle, and leave, based on kinship, family, and friendship. In this way, the reproduction of *family* and *household* are sepa-rate but connected; family, kin, and others may live together under one roof, share various tasks (Eder 1999, 11), and eat from the same pot of rice (Peluso 1990, 171).

The fact that upland families coreside in complex ways (e.g., polygamy, multiple relatives) that fail to align with notions of self-contained households in spatially fixed villages (Macdonald 2007) stands at odds with dominant nonstate (and state) sedentarizing and civilizing reforms. Despite centuries of regional trade and the more recent adoption of lowland agriculture and living—reminiscent of Malkki's (1992, 31–34) "Sedentarist Metaphysic" —highland family groupings arguably retain degrees of sociospatial mobility (Warren 2007; Macdonald 2007). Not all highland families aspire to permanently settle in lowland villages and adopt the typical associations of fixed-plot cash cropping, surplus profit, and capital accumulation.[6] Nonstate (and state) actors' long-standing objective of encouraging Indigenous uplanders to adopt lowland market logics, commercial agriculture, and sedentary living are complicated by fluid Indigenous ontologies embedded in highland life, livelihoods, and landscapes—shifting forms of personhood, moral obligations, reciprocity, ritual, and livelihood across human and nonhuman relations challenge rigid and fixed social categories (Macdonald 2007; Smith 2015; Theriault 2017; Dressler et al. 2018).[7] Biopolitical reforms are no easy feat.

Despite such obstacles, nonstate actors persist in their endeavors. They reaffirm that Pala'wan and similar groups need "help" because they are poor, vulnerable, and destitute—they exist without adequate state protections (Brown 1991; Macdonald 2007; Smith 2015; Theriault 2017). Pala'wan continue to be represented as unruly and ungoverned peoples in need of fixity, discipline, and reform (Macdonald 2007).[8]

Materialities

Nonstate reform programs have long harnessed socially situated material objects to mediate and maximize the effectiveness of their interventions, particularly those items that most matter to Indigenous peoples' ways of life (Bovensiepen and Rosa 2016). The most symbolically meaningful and valued forms of material culture are often controlled and manipulated to reform other facets of Indigenous life and livelihood that are deeply intertwined with such objects. Missionaries and conservation NGOs well know that many customary objects and practices inform familial relations and livelihoods, and that reforming them can significantly influence social reproduction over time. Many also understand that the vitality of material objects in Indigenous lives can mediate and refract their reform agendas. I elaborate on this relationality below.

Socially Situated Objects

In his classic text *The Social Life of Things*, Appadurai (1996, 5) writes, "We have to follow the things themselves, for their meanings are inscribed in their forms, their uses, their trajectories. It is only through the analysis of these trajectories that we can interpret the human transactions and calculations that enliven things." Objects—in this case, commodities—take on social meaning, value, and agency through social and economic exchange, and through the context of the people and places in which exchange unfold. Beyond the bare form and function of exchange, the social value, meaning, and character of objects emerge through relational processes of production and exchange over time and space. Analyzing the social history and contemporary meaning of an object among different peoples and places reveals its changing cultural character, economic values, political potency, and varied ecologies.

As objects move through, take on, and reproduce the social substance of places, they will take on social lives themselves—they constitute a partial biography of upland peoples' lives, livelihoods, and struggles. As Hoskins (1998, 7) notes, the "lines between persons and things can blur and shift . . . [and] objects (cloth, jewelry, porcelain dishes) are often endowed with the qualities of persons," families, and communities. Among Pala'wan uplanders, the "more introspective, intimate, and personal accounts of many peoples' lives [emerge] when asking them about important objects central to personhood and making a living" (2). The relationships between people and things are often less than obvious. Even mundane objects seemingly relegated to the "cultural background" embed within and influence the social lives of uplanders (Miller 2005). The tingkep is one such item that habituates and prompts the lives and livelihoods of the Pala'wan.

The tingkep and other customary items are used symbolically in ritual, livelihood, and everyday practice, reproducing cultural meanings through human and nonhuman worlds. The production, use, and presence of such objects weaves together various elements of Indigenous personhood, identity, social relations, and belonging amid changing situations and circumstances. The cultural roles and positions of customary objects (e.g., their use in rituals) transcend, but are also supported by, their pragmatic roles in everyday tasks and functions (e.g., storing clothing, surplus rice, or cassava tubers), dissolving clear-cut distinctions between objects of sanctity and utility. Because the use and presence of customary objects "materialises [in] social relations, desires and values" (Bell and Geismar 2009, 3), the significance of such objects frequently shifts from the extraordinary to the ordinary in household and community settings. Customary objects work as complex vehicles of meanings, values, and practices in the reproduction of social life in the uplands (Miller 2005).

Mediating Practices of Reform

Centuries of missionary proselytization—in partnership with or enabled by state authority—have targeted and disabled customary objects and practices central to Indigenous peoples' lives through the sustained imposition of new beliefs and practices. In his pioneering book *Christian Moderns*, Keane (2007, 5) explores how missionary endeavors undertook the work of "purification" to delegitimize and control Indigenous material practices and support their "liberation from a host of false beliefs and fetishisms that undermine freedom." Acts of purification encompassed sustained efforts to *dematerialize* Indigenous "signifying practices" so as to surmount cultural constraints on autonomy and freedom, the basis of "modernity." Keane illustrates how the Protestant neoorthodox Reform Church and Christian Church attempted to reform Sumbese beliefs in eastern Indonesia through the sustained deconstruction and conversion of their language, signs, and practices—that is, the gradual dismantling of the semiotics of Sumbese material culture and ritual practices (Keane 2007). Protestant missionaries were typically reluctant to accommodate Indigenous cultural forms and expressions of Christianity (Hefner 1993).[9]

The missionary desire to "drive a clear line between humans and nonhumans, between the world of agency and that of natural determinism" sustained itself across the archipelagic region (Keane 2007, 7). In 1908, Catholic missionaries in Timor carried out deeply violent reforms by attempting to burn ancestral *lulik* houses. The missionaries believed that the sacred powers of these objects were "the major obstacle to their embracing the Christian religion, and that only through the total destruction of such "fetishized" objects could the "souls" of the Timorese be saved" (Bovensiepen and Rosa 2016, 664). In contrast, Schiller (2009, 281) describes how Catholic missions of the nineteenth century encouraged the mixing of *adat* and Catholicism among Dayak groups in Borneo, but that the Protestant Dayak majority was subject to stricter interpretations of faith under the more conservative *Gospel of the Tabernacle*. Protestant missions usually forbade the use of customary languages, dance, and dress in church services. Despite these distinctions, however, both Catholic and Protestant missions strove to suppress the social and material meanings of adat by implementing formal schooling and other Christian strictures in the remote forest settings of insular Southeast Asia.

Missionaries applied similar biopolitical logics to Indigenous material culture and social practices across the Philippines (Howell 2009). McCoy (1982, 155) describes how the missionaries under early Spanish rule (1600s–1700s) regarded the "animism [*sic*]" of Filipinos and the *babalyan* (*beljan* in Pala'wan) in particular as a "credible evil to be confronted and overwhelmed with the might of Latin invocations and Christian symbols." Catholic missionaries of the time believed that their sustained proselytization had removed the "palpable manifestation of Satan's presence" among local peasants in the Tagalog region who had apparently enthusiastically embraced the "preaching of the holy gospel" (155). As chapter 4 details, many missionaries during the Spanish and American (1898–1946) colonial period focused incessantly on prohibiting the babalyan's use of supernatural talismans—the *anting-anting* (*mutja* in Pala'wan)—often carried in coat pockets, sachets, or tingkep. The anting-anting was known to bestow superhuman powers and courage upon the user, requiring that missionaries isolate and suppress the talisman's mystical potency from the rationality, piety, and morality of Christianity (Ileto 1979; McCoy 1982; Wheatley 2018). Missionary reform initiatives thus aimed to separate human agency (subjectivity) from customary beliefs and objects (materiality) to realign peoples' conception of agency with the individual Self, rationality, and greater freedoms (Keane 2007). Despite their persistent interventions, many missionaries soon realized that the effects of their proselytization were largely superficial, with many peasants retaining or returning to ancestral customs.[10]

Despite contrasting objectives, conservation NGOs have instituted similar longer-term trajectories of reform. However, rather than denigrating and suppressing customary objects and practices, they often try to valorize the value and material basis of certain objects so they can be reproduced (e.g., planted, woven, sold) to incentivize behavioral changes and overcome the supposedly destructive practices often associated with such objects. Here, the ontological question of biopolitics—*what is allowed to flourish and what is allowed to die*—is on full display. Like missionaries, such reforms invoke grand narratives of modernity, while simultaneously valorizing and subjugating Indigenous peoples through sustained interventions that specialize in realigning customs, objects, and livelihoods towards specific end-goals. In valorizing the production and exchange of customary materials and practices—whether Indigenous baskets or agroforestry fallows—the reframing of an object's social and material character aims to reproduce enough social and economic value to commodify, manage, and modify forest-based livelihoods. In turn, customary objects and livelihoods are recast as market-oriented, authentically Indigenous, and seemingly sustainable.

Biopolitical reforms seldom immediately or completely replace Indigenous social and material relations, however. Sociomaterial objects and practices have "tenacity" formed through the social and physical processes from which they emerge, within and beyond highland societies. Such staying power has partly to do with how both social and material processes are coconstituted as cultural products with meanings, significance, and power that are at once deeply recognizable and significant, as well as subject to reinterpretation and modification

(Bennett 2004; Ingold 2007). In this sense, customary objects and practices invariably mediate upland life and biopolitical practices as they coemerge in line with changing beliefs, needs, and desires across places (Bennett 2010). Such items are forged relationally and have the "socioculturally [and biophysically] mediated capacity to act"—or, more precisely, degrees of agency (Ahearn 2001, 12). In Bennett's (2004, 348) words, "cultural forms are themselves material assemblages that *resist*."

An object's agency stems from the social actions and material characteristics that emerge as it assumes and influences the identities, meanings, values, and functions of the people and places through which it moves (Hoskins 2006; Miller 2005). The agency of material things emerges from those who construct it, who interact with it, who use it, or who possess it in contrasting settings (Hoskins 1998, 2006). The tingkep's power emerges as it takes on and shapes the personality and intentionality of its makers and users—the weaver, the spirit, the buyer, or the seller—and the character of livelihoods and forest ecologies across human-nonhuman relations. The basket and its properties (size, weave patterns, form, and associated spirits) are entangled in and affect social and ecological processes within and beyond upland settings.[11] Customary objects, beliefs, and livelihood practices not only coconstitute Indigenous worlds but also challenge reform projects with the aim of overcoming them.

Yet overemphasizing the agentive capacity of sociomaterial objects and practices runs the risk of essentialization. Material cultures, whether Indigenous or otherwise, are anything but absolute, static, or time-bound. Like most humans, Indigenous uplanders define, give form to, and value customary objects and relations within the context of changing political and economic conditions. Far from being averse to commerce, they themselves have facilitated the objectification and alienation of customary objects as commodities to sell in markets near and far for centuries. Although seldom (if ever) fully alienated from social relations (Cohen 1989), customary objects have long been commodified through monetary exchange and value as they traveled from hinterlands to regional entrepôts (see chapter 3; Warren 2007).

Depending on the character of market relations, objects may gradually or quickly take on greater monetary value, disembed from social relations, and assume new values along circuits of production and exchange (Appadurai 1996; Miller 2005). However, even when those who produce customary objects are alienated from their social substance (e.g., reciprocal relations, obligation); livelihood function; and broader ontological connections (Theriault 2017), an object's meaning, value, and agency are not simply lost through its commodification (Graeber 2001). It takes on new forms, meanings, and functions across space. In the northern Philippines, for example, Ifugao women's craft production drew them deeper into capitalist relations with various negative implications (Milgram 2001). However, they found ways to work together to cooperatively showcase, value, and sell their products, overcoming the precarity of market fluctuations (Grimes and Milgram 2000; McKay and Perez 2017). Across the Philippines, Indigenous weavers and nonstate actors have creatively capitalized on the growing recognition of their identities, customs, and traditions, often reassembled as Indigeneity, to valorize and "authenticate" customary objects. By leveraging culture and custom, they further legitimate their beliefs, livelihoods, and land rights, often irrespective of whether such markers of Indigeneity reflect local realities. Nevertheless, should Indigenous peoples lose control over how their material culture is produced, exchanged, and represented, the impacts and outcomes can be adversely consequential. In commodity form, a customary object is easily fetishized and transformed to fulfil the desires of consumers—whether NGOs or tourists—such that the social history and ontological basis of the object is masked and rendered invisible (Graeber 2001), violating Indigenous rights, beliefs, and heritage. Ultimately, the fetishizing of customary objects and livelihood practices in order to overcome them reinforces the production of social difference between those who seek reform and their biopolitical subjects.

Indigeneity and the (Bio)Politics of Difference

In Southeast Asia, the substance of state and nonstate biopolitical practices have origins in racially normalized categories, representations, and orderings that extend from the colonial to postcolonial period (Scott 1974; Li 2007; Blanco 2009; Scott 2009). Historical records show in detail how colonial and religious authorities have constructed and essentialized upland peoples based on what they overcame, dispensed with, or never found (Scott 1974; Wolf 1982). Such representations of what was and or what became of Indigenous peoples were based in assumptions that these peoples' characteristics were somehow evidently different from those of colonizers—that is, based on their apparent social and cultural distinctiveness. Such representations cast upland societies in terms of their cultural structures, that, unlike those of "modern" European society, supposedly maintained a coherent, homogenous cultural identity (see chaps. 4 and 5). As Wolf (1982) recognized, such assumed cultural differences were abstracted out as coherent wholes, or "cultural things," whereby Indigenous societies came to be seen as apolitical and ahistorical entities, represented as a "people without history."

In postcolonial contexts, Indigenous uplanders continue to be constructed as recalcitrant, primitive, and remote peoples, threatening to those in power despite or because of their spatial distance. Such politically reified cultural differences morally obligate those in power to reform unruly tribals into modern compliant subjects (Li 2007; Keane 2007; Scott 2009). Today, nonstate actors assist, and ultimately bypass, governments in the discursive production of cultural difference through local-to-global imaginaries of reform. To them, and to some uplanders, the articulation of difference consciously reproduces coherent cultural groupings and ethnolinguistic wholes that underpin notions of Indigeneity. In time, emerging global imaginaries of Indigeneity produce and fill what Li (2000) aptly called the "tribal slot": a discursively bounded political category that neglects uplanders' changing identities, aspirations, and practices (Eder 2013).

State and nonstate actors continue to represent highland peoples through narrow socioecological repertoires, naturalized representations that typically permeate broader society as racialized tropes. Fixed and enduring sociocultural traits and objects (customary items, practices), physical characteristics (darker skin, lean bodies), traditional livelihoods (hunting, swidden), length of occupancy (long term, since time immemorial) and other essentialized traits define and steer interventions toward them. Indigenous people may draw on such "emblems of difference" (Barth 1998) to reaffirm their own identity, belonging, and rights but, in doing so, may further marginalize themselves by amplifying social difference (Li 2000). The reification of cultural difference—indigeneity (*katutubo* or Pala'wan) or otherwise—is thus a double-edged sociopolitical process. For many nonstate actors, the notion and substance of Indigeneity has become a beacon of reform and optimization, a means of achieving biopolitical ends and modern ideals. For Indigenous peoples themselves, it can be a source of pride and resistance as well as condescension and shame (see chaps. 6 and 7).

Becoming an Indigenous Uplander

The notion of Indigeneity—a discursively constructed category, lived reality and social practice—is informed by racialized colonial ideals, contemporary human rights, and global environmentalism in the Philippines and elsewhere in Southeast Asia. The broader category of Indigeneity often "impl[ies] first-order connections between a group and locality. It connotes belonging and originariness and deeply felt processes of attachment and identification, and thus distinguishes 'natives' from others" (Merlan, 2009, 304).[12] Despite many Filipinos having some direct, sustained connection to place, language and kin, in the late 1990s the Philippine state, along with nonstate actors, established laws, policies, and practices that interpreted Indigeneity as comprising organic first-order origins (e.g., demonstrating "tribal" ancestral lineage, ritual, ethnolinguistic homogeneity) and degrees of political marginalization (e.g., historical oppression, discrimination). Heavily influenced by global ideals and national sentiments, such politicized cultural categories have been sutured and repurposed in civil law and local practice to render "Indigenous Filipinos" distinct from dominant Filipino culture and society (Dressler 2009; McDermott 2001; Theriault 2019).

In the Philippines, the recognition and substance of Indigeneity are neither natural nor inevitable (Scott 1974; Li 2000). Rather, the notion was forged early on by colonial actors who produced and conflated a racialized tribal character with upland peoples who, in being codified by such labels, were naturalized as tribal Others. Historian William Henry Scott (1974) and others have suggested that under the workings of Spanish and American colonization, administrative policies and practices constructed racialized non-Christian tribes that would later be reclassified as "tribal peoples," "Indigenous cultural communities," and, finally, "Indigenous peoples" needing modern rights and emancipation in post-colonial contexts. As chapters 3 and 4 show, such colonial constructions and contemporary imaginaries of the tribal Other permeate biopolitical programs throughout the Philippines, informing how both state and nonstate actors characterize, codify, and manage Indigenous uplanders, their lands, and their livelihoods (for a discussion about adat in Indonesia, see Li 2000).

Since the 1990s, then, nonstate actors have increasingly invoked the spectacle of the tribal Other to legitimize their reform agendas. In Tsing's (1999, 159) words, the fantasy of "becoming a tribal elder" has persisted as a dream machine that reproduces "the rural, the backward, and the exotic." However, as missionaries, some NGOs and state actors still speak of optimizing "tribal peoples," long-running peasant and Indigenous rights struggles in the Philippines (and elsewhere) have produced positive outcomes with the assistance of progressive actors in civil society and government, leading to legal and political recognition of Indigenous rights, identities, and claims to lands, forests, and waters. Often, these hard-won struggles have successfully sought redress, healing, and reconciliation from centuries of deep injustices caused by the violence of colonial and postcolonial governments, conservation organizations, religious institutions, militaries, and extractivists (Kuper 2003; Niezen 2003).[13] As a contemporary political project, the performance of indigeneity can be both essentializing and emancipatory as it works through conservation and human rights imaginaries across scale.

Global Indigenism and Ecological Labor

New alliances emerged in the 2000s between variously positioned environmental NGOs and Indigenous peoples' organizations that centered on enrolling Indigenous labor for brokering and managing forest and land management projects in remote highland areas. However, the assumptions informing such partnerships and the agreements themselves, were most often bound by uneven social contracts, advancing narrow governance objectives, and ultimately sidelining Indigenous peoples' control over land and forest resources (Chapin 2004; Niezen 2003; Igoe 2005; Dressler et al. 2010).

Rather than necessarily aiming to overcome all things "customary"—as missionaries often do—state and nonstate forest governance ideals have returned to valuing and reifying "Indigenous attributes" in ways that have legally compelled Indigenous subjects to uphold and conform to them with little justification and regard for their changing realities (Povinelli 2002).[14] The presumed authenticity and intactness of tribal features (e.g., sustained traditional ecological knowledge, sustainable Indigenous land management, and light extractive practices) continue to be valorized as seductive biopolitical drawcards. In the context of "collaborative management," many Indigenous peoples including the Pala'wan find themselves negotiating what Conklin and Graham (1995, 695) call an ever-shifting "middle-ground" of contested knowledges, identities, and practices. In this political space, Indigenous peoples and nonstate actors populate a mutually intelligible but conflictual space to strategically draw upon each other's symbolic ideas, imageries, and agendas for political ends (696).[15] Yet the interpretations of mutual expectations, the quality and duration of benefits, and the type and amount of labor provided is uncertain (Neimark et al. 2020; Greenleaf 2021).

As nonstate actors implement upland projects they will often draw on Indigenous peoples' ecological (or religious) labor—the affective and performative physical, mental, and psychological care work underlying programs and projects—to sustain and regulate biopolitical interventions. Bound up in benefit-based agreements, often-precarious Indigenous labor becomes further racialized and naturalized as necessary for successful collaboration and project completion (DiNovelli-Lang and Hébert 2018; Neimark et al. 2020; Greenleaf 2021). However, the appropriate social valuing and financial compensation for relying on Indigenous labor for "preventing, repairing and mitigating ecological degradation" is often lacking in contractual, collaborative initiatives, whether with environmental NGOs or evangelical missions (DiNovelli-Lang and Hébert 2018, 1). Both groups will draw on and leverage Indigenous labor for their respective biopolitical enterprise, without sustaining the long-term benefits expected by uplanders who forgo beliefs and forest use in exchange (Greenleaf 2021).

As a result, these unstable alliances ultimately collapse when one actor deviates from agreed-upon roles and representations. Such outcomes occur most frequently when ill-conceived project interventions reproduce fictive realities that spectacularly misalign with the lived realities, experiences, and expectations of Indigenous peoples. Despite the more respectful "politics of difference," nonstate (and state) actors continue to conflate otherwise fluid notions of Indigeneity

with static and fixed tribal markers, along with moral imperatives of both purification and optimization (Brosius 1999; Kuper 2003). Contrasting realities thus
impinge upon mutual expectations, often with deleterious outcomes.

Practices of biopolitical reform invariably intersect with and influence how
Indigenous peoples live and make a living in the forested uplands of the Philippines and elsewhere in Southeast Asia. Although powerfully mediated by changing political interests, aspirations, and spirit worlds (Rubis and Theriault 2019),
the influence of emplaced nongovernmental reforms finds ways to work within
and through processes of social reproduction, livelihood, materiality, and self-
identification among uplanders over time and space. This conceptual overview
sheds some light on how historical and contemporary practices of state and
nonstate reform shape and are shaped by everyday life and livelihood among
Pala'wan uplanders. Pala'wan sociomaterial relations, livelihoods, and indigeneity are historically contingent and translocal expressions of social difference,
survival, and, ultimately, resistance. Yet, in the face of reforms, how they live and
reproduce themselves is increasingly subject to the violence of biopolitical rule
and global environmental imaginaries.

Framed by biopolitics and conjuncture, the subconcepts of social reproduction, sociomaterialities, and the politics of difference can, when taken together,
further elucidate how nonstate actors have reified Indigenous highlanders as
objects to valorize, overcome, and, ultimately, reform through governmental
projects of rule. Building on colonial legacies, NGOs celebrate Pala'wan as the
authentic, morally pure Filipino embedded in nature—with authentic rituals
and material cultures—whose citizenship matters little as customary practices
are expunged in the name of Jesus and modernity (Casumbal-Salazar 2015). The
remainder of this book explores in detail how these racialized governance ideals,
(mis)representations and localized impacts have unfolded over several centuries
in the highlands of southern Palawan.

UPLAND LIVING AND TINGKEP WORLDS

> Our uma provides us with rice, so we do not have to buy rice in the lowlands. It's in this area that we were born with our culture and ceremonies, like lutlut. The rice we cook in the bamboo [shoots] must come from our own uma, not from rice paddies down below.
>
> —Baltazar Binag, Marenshewan, Bataraza

For many Pala'wan, the uplands have long nourished social relations, customary practice, and livelihoods largely independent of the state. The uplands are territories of life and livelihood (Escobar 2020)—moist forests, heirloom seeds, ecological knowledge, deities, and family labor coproduce swidden mosaics valued for social and physical nourishment. As Baltazar explains, culture and ceremony require swidden rice, not less desirable wet rice from the lowlands. Similarly, many Pala'wan weavers describe how human and nonhuman relations constitute social and material practices in the forested uplands. Forest animals influence designs on life and how to remember it. As elderly weaver Nelito Tipack explains: "We weave patterns from the *mata puney*, the eye of the morning dove. . . . We learned this from our ancestors and those dwelling in the forest." Diverse forests and their inhabitants have long informed designs evident in the social life, materiality, and livelihoods of ancestral landscapes. It is the uplands, not the lowlands, that are recognized as ancestral home.

This chapter offers a window into Pala'wan social life, customs, and livelihoods along the southeastern regions of the Mount Mantalingahan Range. I position the social and cultural character of customary objects—particularly the tingkep—in relation to the rituals, myths, and forest livelihoods that intersect with the corporeal and spiritual realms of Pala'wan in and near Kamantian and Marenshewan. This background frames how Pala'wan existence articulates with projects of reform explored later in the book. My partial account of Pala'wan lifeways sheds light on how and why they contend with nonstate actors attempting to undo what has taken centuries to form: a relatively autonomous, self-sufficient, and defiant existence.

The Pala'wan

The Pala'wan are one of three main Indigenous groups on Palawan Island. They number more than fifty thousand and predominantly reside on state lands (public domain or timberlands) in the forested mountains, valleys, and, increasingly, the lowlands of southern Palawan (Macdonald 2007). They speak an Austronesian language that is generally unintelligible to other Indigenous uplanders like the Tagbanua and Batak. Pala'wan subgroups—each with varied dialects, rituals, material culture, and beliefs—are named after the rivers, watersheds, and other sociomaterial features marking their location in the mountains of southern Pala'wan (Revel 1990; Macdonald 2007).

Pala'wan subgroups are often sociospatially differentiated: Pala'wan 't daya/ bukid are people of upstream or mountain areas, while Pala'wan 't napan live downstream, in lowland areas, plains, and coasts (Revel 1990, 84–85; Novellino 2001, 78). Upland swidden practices, forest-based livelihoods, and social relations spatially reflect complex subsistence-cosmological orientations along this "inland-coastal gradient" (Macdonald 2007, 12, 33). Lowland areas tend toward fixed, legible, and readily governable private land holdings, cash cropping, commercial activities, and monotheistic practices. However, these upland-lowland gradients are beginning to blur. Increasingly, highland Pala'wan engage with Barangay politics, migrant (and other Pala'wan) landholders, and merchants to diversify their livelihoods through the social and economic opportunities of the agrarian lowlands. Declining swidden rice yields and fewer nontimber forest products (NTFPs) because of restrictions on forest access, difficulties hunting game, and growing indebtedness compel Pala'wan to seek wage labor on lowland paddy rice farms (basakan); expanding (copra, rubber, and palm oil) plantations; coastal fishing pursuits; construction sites; and in nonstate programs (Eder 1999; Dressler and Fabinyi 2011).

Most upland Pala'wan families nevertheless continue to draw on a socioecologically complex suite of livelihood practices—including, but not limited to, swidden agriculture, diverse tree and root crop cultivation, and the harvesting of NTFPs and riverine species. These activities are supplemented by cash from wage labor or trading/selling upland goods (e.g., cassava for fish and salt) in the lowlands (Macdonald 1992a, 1992b, 2007; Smith 2015; Theriault 2017). With greater elevation and distance from town centers, Pala'wan tend to be more reliant on, and self-sufficient through, the diversity of flora and fauna in lush ancestral forests. This combination of pliable, resilient crops, mobile swidden fields, and acephalous residency patterns have made these and other highlanders historically less governable by state authority (Scott 2009). This independence is in decline with the increasing reach of nonstate actors.

Until relatively recently, state-supported infrastructure development in southern Palawan was confined to intensive irrigated rice and copra production in lowland areas. State governance and influence were relatively limited on the interior fringes of Pala'wan territory until the mid-twentieth century. The influence of the Catholic Church during the Spanish colonial period was also limited in the Palawan uplands. As Macdonald (1992a, 128) notes, "The Pala'wan were only superficially exposed . . . to Catholicism. There was . . . no systematic attempt at Christianisation," beyond small parishes in the lowland towns of Brooke's Point and Quezon. In contrast, Islam influenced Pala'wan beliefs and practices for at least six centuries through proximity and trade with the Sulu Archipelago and Sultanate (Macdonald 2007; Warren 2007).

Like most uplanders in Southeast Asia, Pala'wan beliefs and practices have long engaged with capitalist relations that intersect and bridge upland and lowland spaces. Uneven agrarian market relations and political economies established by colonial and central administrations and largely under the control of migrant-settlers have marginalized upland Pala'wan, relegating them to subordinate positions with less authority and control over ancestral lands and forests. They have limited bargaining power over commodity prices and labor conditions or over who enters and remains in the forested uplands (Brown 1991; Macdonald 1997). While some lowland Pala'wan families have successfully negotiated the local agrarian political economy—their children finished high school and college and took up work in the extractive sector, conservation, tourism, or government—most Pala'wan remain steadfast swidden farmers, hunters, and NTFP harvesters in a context of uneven agrarian change. They attempt to skillfully combine diverse self-sufficient upland livelihoods with wage labor to meet their income needs amid declining yields and rising debt.

Sustained biopolitical reforms by environmental NGOs and Adventist missions have reinforced this uneven social, economic, and environmental change and dismantled self-sufficient ways of life in an unsettling bid to optimize the Pala'wan by eradicating swidden, sedentarizing livelihoods, and aligning their beliefs with Christianity. The numerous adverse and interrelated outcomes of these reforms include the erasure of beliefs, the emptying of forests, the transformation of livelihoods, the degradation of forest landscapes, and, ultimately, the loss of political autonomy.

In this chapter, I broadly focus on the "cultural ecology" of Pala'wan living in the Mekagwaq-Tamlang watershed (lit. place of "sad solitude"; Revel 1990a, b;

FIGURE 2.1.   Pala'wan households in the highlands, 2013. Photo: Dressler.

see fig. 2.1) and the more southern Marangas region of Bataraza before discussing their responses to varied biopolitical reforms (Revel 1990b). I explore the diverse livelihood strategies of Pala'wan families who, until relatively recently, sat somewhat beyond nonstate influence. I position *uma*-making (swidden) as the basis of familial social reproduction and independence and consider how the materiality of the uplands, particularly the tingkep, comes to mediate Pala'wan lives and diverse nonstate reforms.

Ungovernable? Pala'wan Social Relations, Adat, and Residence

As Macdonald (2007, 91–92) observes, the Pala'wan have "no unified canon of religious law and beliefs and practices vary" across interior watershed areas. Pala'wan beliefs and practices are best described as being passed down by *Adat et Kagunggurangan* (lit. "the tradition of the Ancestors"; Revel 1998, 7; Macdonald 2007, 2009). Adat, or customary law, comprises rules of etiquette and judiciary that most Pala'wan groups uphold (Revel 1990; Macdonald 2007, 2009), particularly in the interior (Macdonald 2007). Adat is a form of oral litigation, exercised under the guidance of respected elders with expertise (*megsumbung, memitsara,* or *panglima*) in arbitrating lengthy oral discussions or "public hearings" (*bitsara*). These discussions may involve litigation, deliberation, and negotiation concerning marital affairs, theft, and other social issues. An elder panglima mediates the proceedings in line with customary law (87). The outcome depends on the social esteem and respect for the arbiters and the potential for collective decisions since, in "all cases the final discussion will have to be agreed upon by both parties. . . . Therefore, a real consensus has to be produced" (87).

Western religious concepts and terms like "sacred" and "priest" are less appropriate for upland Pala'wan than for settlers in the lowlands. Nonetheless, Pala'wan beliefs saturate human and nonhuman worlds with degrees of consistency—sometimes differing only slightly from other Indigenous peoples on the island—resulting in a regularized system of beliefs that loosely constitutes a form of religion.[1] The predominantly male ritual specialist (on Palawan), the beljan, serves as a conduit for these complex beliefs, roles, and functions across human and nonhuman worlds. The beljan is a polymath ritual specialist—a healer, arbiter, counsel, intermediary, and medium—who assists Pala'wan families and communities in contending with uncertainty across multiple human and nonhuman realms (see chap. 6). The beljan and other Pala'wan figures of customary importance are afforded esteem and respect, but not revered as divine, religious figures (Macdonald 2007, 91).[2] During my conversations with the few remaining beljan, however, many describe, with some reverence, the all-powerful, all-seeing Tungkol, a highly respected beljan whose great powers can cure many illnesses, revive those who have recently died, and see deep into the spirit world. They are considered immortal and part of the highest plain of the Pala'wan universe, the realm of *diwata* and *taw kewasa* ("the powerful ones"; see below and Macdonald 2007).

In recent decades, many respected Pala'wan have become entangled in state and nonstate politics of indigeneity and, specifically, Indigenous rights to land and resources, knowledge, and biodiversity conservation (McDermott 2000; Dressler 2009; Theriault 2019). In the process, state and nonstate actors often impose—and some Pala'wan eagerly accept—rigid leadership titles or categories upon otherwise fluid Pala'wan social structures that loosely correspond with lowland political structures. In this mix, it is nongovernmental entities that explicitly target customary leaders' legitimacy and authority to advance their biopolitical objectives across the southern highlands—missionaries aim to overcome their authority, while environmental NGOs aim to valorize it.

Environmental NGOs use customary leaders in varied bureaucratic initiatives, ancestral domain delineation, and livelihood projects across the island's forested uplands (Smith and Dressler 2017; Theriault 2017, 2019). Pala'wan leaders may work as local brokers and translators for programs that aim to curb swidden (e.g., through fixed-plot agriculture), encourage increased handicraft production to generate more income (to offset the need for hunting and clearing forests for food), and facilitate reforestation programs in the uplands. These leaders' titles and roles harden politically as they perform them among their own and become imbued with greater authority in their communities. The political influence of a newly forged "tribal chieftain" may even extend outward to other villages and different NGOs.

In contrast, missionaries aim to overcome and replace the customary roles of Pala'wan who broker human-nonhuman relations, particularly the beljan and panglima. To effectively halt the social reproduction of adat, myths, and rituals, missionaries explicitly target and reform those individuals whom they see as (in a functionalist sense) the "cultural core" or "source" of customary beliefs. They even draw young Pala'wan children away from their families to convert them in evangelical high schools and churches (see chap. 4). Once custom is stripped, they then endeavor to cultivate pious Pala'wan as stalwart religious leaders.

Anarchistic Social Relations

Despite the rigid titles and hierarchies imposed by state and nonstate initiatives, most highland Pala'wan continue to invest in social relations, organization, and residence patterns that remain broadly nonstructural, acephalous ("headless") and largely anarchic (Macdonald 2009, 2011; Revel 1998). The relative lack of leadership hierarchies and authority to command exists alongside strong principles of sharing and equity (Revel 1998; Macdonald 2007). Any notion of authority among Pala'wan is best understood as the capacity to influence or persuade through kinship roles and expertise in customary law (Macdonald 2007). Social principles of sharing (particularly game meat and rice), cooperation, and reciprocal exchange are upheld as morally appropriate behaviors based in custom and survival. As Revel (1998, 1) writes, with a tinge of romanticism, "Palawan ideology is profoundly peaceful and egalitarian. The values of cooperation in work, *tabang*, sharing, *bagi*, and equal exchange, *gantiq*, form the basis for relations with others and the world. . . . Life in the forest demands altruistic and group-oriented behaviour."

As with other upland peoples, however, these social relations are increasingly punctuated by a growing recognition of private property, formal ownership, accumulation of surplus, and the need to generate a sustained income—for example, from selling pig meat or forest honey at *sari-sari* stores (small general stores) or selling land for cash and/or palm oil production. Lowland agrarian political economies, marked by varied reforms, have further entrenched Pala'wan into uneven capitalist relations.

Residential Patterns and Fellowship

Pala'wan residence patterns remain antithetical to nonstate (and state) reforms advocating for fixity and permanence. Small household clusters or "sororal sibling groups" comprising five, ten, or more families characterize Pala'wan uxori- and matrilocal residence patterns. Revel (1998, 1) notes that "the hamlet of . . . families is made up of a group of sisters and first cousins assimilated to sisters, around whom the husbands congregate." A father, an uncle, or a senior figure with customary influence presides over the hamlet to ensure that degrees of harmony are maintained (Macdonald 2009). Household aggregations vary over time—families and individuals come, settle, and leave in a fluid manner based on bilateral kin lines and varied friendships. Old friends and distant cousins may visit for several days or stay for long enough that a room or small house is made available for them. Kinship and acquaintance forge friendships and fellowships that acknowledge shared social ties, relations, customs, and ways of life connected to ancestral forests (Macdonald 2009; Dressler et al. 2018).

Pala'wan generally live in household clusters, forming small-to-medium social units, hamlets, or ("unregistered") temporary settlements that are often disbanded and reformed elsewhere after two to three years (Macdonald 2009, 8). Family mobility, partly influenced by the swidden fallow cycle, is central to sustaining social relations, replenishing forests and soils, and revitalizing crop yields. In the highlands, most Pala'wan hold land under family usufruct based on occupancy, use, and the labor invested in planting and clearing. Planting permanent tree crops (such as fruit trees) often marks a more enduring claim to land, although the claim attenuates as fallowed land returns to the forest commons (Macdonald pers. comm. 2020; see also McDermott 2000; Dressler 2009). The uma field and crops may move with family and kin to more favorable upland locations, where land is more fertile and political threats less apparent. The mobility of familial livelihoods allows for the possibility of retreat and greater independence, though increasingly less so.

Less-than-Fixed Gender Relations

Pala'wan household reproduction is similarly fluid and involves a relatively symmetrical gendered division of labor, cooperation, and reciprocity across conjugal, familial, and extended kin relations. Ethnographers have previously argued that men and women specialize in complementary social practices (Macdonald 2009, 8). Women typically cook, weed the uma fields, glean, and weave; men hunt, fell trees in umas, and construct homes. However, these social boundaries between women's and men's activities in the home, forest, and field are not fixed (Macdonald 2007, 67). My own observations suggest that Pala'wan gender roles are fluid and often reversible, as Pala'wan women and men mix and share labor requirements based on necessity. For example, when labor shortages arise during uma preparation, the typically gendered roles of weeding and seeding rice are reversed. Men take on the women's task of weeding, while women prepare the ground using dibble sticks for men to plant rice seeds. Revel (1990, 149) describes a firmer gendered division of labor concerning forest materials: the work of Pala'wan women typically involves the softer, more flexible forest materials (e.g., stripped, soaked pandanus or softened strips of bamboo and wood pith) used to weave baskets and winnows, while the work of men involves the harder forest materials (dense hardwoods) used to carve statues, axe handles, or housing materials. However, I again found that the opposite also holds true: Pala'wan men often weave tingkep while women fell round timber.

Nonstate reforms often aim to naturalize gendered divisions of labor. Missionaries can seek to reify the status, authority, and control of male household "heads," and environmental NGOs frequently broker social contracts with male household members while relegating women to the role of "cobeneficiaries." Nonstate actors also often seek to reform the anarchic social relations of Pala'wan as part of sedentary "civilizing missions" that align with long-standing state governance mechanisms that encouraged fixed patterns of tenure and livelihoods.

In the midlands and lowlands, Pala'wan households increasingly use permanent tree crops, fences, and other geographical markers, like rivers, to make semiformal claims to land and anticipate claim recognition through a so-called tax dec (i.e., a paper tax declaration receipt). This receipt, issued by land surveyors who happen to travel beyond the coastal plains, reflects the state's partial recognition of their de facto claim to land (inside or outside Certificates of Ancestral Domain Titles, or CADTs). Many lowland Pala'wan families have also acquired permanent holdings by petitioning the state or by releasing alienable and disposable lands.

Despite these changes, the relative political autonomy of upland Pala'wan existence continues to be upheld through shared identities, language, beliefs, ritual practice (Revel 1990), livelihood, material culture, and memories across human and nonhuman worlds (Macdonald 2007). It is also maintained through the occupancy, use, and transformation of swidden forests, waters, and ancestral lands that the Pala'wan have held since time immemorial (Theriault 2017; Smith 2021). Indeed, in highland areas, Pala'wan households' social relations and proximity to one another produce a form of "fellowship": "an association of friends, of equals who share a common interest and freely choose each other's company to pursue a common interest" (Macdonald, 2009, 9) in the absence of formal government boundaries and administration (Barclay 1996; Macdonald 2009; Scott 2009).

Intersecting and Changing Pala'wan worlds

For the Pala'wan, elements of monotheism coexist with a human, material realm and a nonhuman, invisible realm comprising malevolent (*taw't talun*) and benevolent spirits beings (*diwata*, deities) who dwell within and beyond the visible landscape (Revel 1998; Macdonald 2007; Smith 2015; Theriault 2017). The supreme being, true god, or "Master," Empu Banar (hereafter Empu, also spelled Ampuq), wove the world into fifteen layers: "a universe with seven levels above and seven levels below" the visible world of humans (Revel 2011, 51). Empu lives "in the highest level, *Anduwanän*, whereas the 'earth', *dunyaq*, is the lowest of the levels above. Invisible now in this ultimate abode, he [Empu] is carefully watching over us, the 'genuine human being,' *Taw banar*" in the visible realm (52). While Pala'wan subgroups have contrasting accounts of their cosmology, they all acknowledge that these invisible realms manifest in ways that powerfully influence the visible elements of human and nonhuman worlds—from environmental governance to the forest materials used for tingkep making.

Indeed, most Pala'wan I spoke to—some distance from the lowlands and the missionary camp—believed that Empu oversaw the various masters and deities in the "middle realm" who protected and mediated Pala'wan negotiating changing natural and spiritual phenomena (Macdonald 2007; Revel 2011).[3] As the creator, Empu belongs to, watches over, and is assisted by other invisible entities also commonly called *empu* (Novellino 2001), who function as masters or lords of natural objects (Macdonald 2007). In this realm, Empu often bestows an invisible humanlike soul (*kurudwa*) to species of socioecological importance (e.g., upland rice, wild pigs), referred to as *taw* (*tao*)—literally, "person." This imbues rice plants and wild pigs (*biek talun, Sus ahenobarbus*) with degrees of consciousness, personhood, and a human double in myth, ritual, harvest, or the hunt (see below).

The animals and plants endowed with Kurudwa each have guardian masters. There are many, including the Master of Pigs (Empu et biek), the Master of the River Eels (Empu et kasili), and the Master of Rice (Empu et parey). From the heavens, these masters mediate and influence how Pala'wan live and make a living, using forest species in diverse swidden landscapes. Empu demands that Pala'wan respect the bodies and souls of these species and practice restraint when hunting them or using their environment. As Empu influences the reproduction of animals or plants (e.g., the rice harvest), those who take too much of a plant or animal may be punished with a physical or mental illness (Novellino 2001, 83).[4] Empu's agency partly calibrates Pala'wan social relations with one another, their environment, and their spiritual world.

Revel (1998) elaborates on Pala'wan relations with the nonhuman world as a "generalised hunt" (*la chasse généralisée*) in which humans are the hunter and the hunted. Like other uplanders, Pala'wan believe that certain species have an animal double for forest deities, which shift between human and animal form in dreams and hunts (Theriault 2017). In this sense, just as Pala'wan who hunt wild pigs are also hunting (invisible) humans who may appear during the chase, other animals (e.g., white chickens) holding human souls are preyed upon by malevolent forest deities such as *taw't talun* (Macdonald 2007; Theriault 2017; Smith 2021). Moreover, the Empu of rice and wild pigs and other significant flora and fauna are known to give what they occupy to those Pala'wan who pursue them and provide a ritual offering of thanks or appeasement. Any inappropriate behavior toward either species—either excessive harvesting or speaking immodestly—is a sign of disrespect that may anger the respective Empu and lead to the refusal of future bounty (Theriault 2017; Smith 2021). Even as Empu offer the desired bounty, the farmer or hunter can also be preyed upon by the taw't talun. Revel (1998, 8) describes this reciprocal chase and ethic of care in greater detail:

> The plants, animals, game, and fish that they eat also have a physical body and an invisible "Master," *Ampuq ya*, which must always be taken

into account. The Master is to natural objects what the double is to the body of the true man. In order to survive in nature, their source of sustenance, men must of course "take"—but not in excess. They can hunt, fish, and cut down forest trees if they make joint decisions and show care and respect by appeasing and making offerings to these other humanities, invisible and omnipresent, which demand that equitable relations be maintained. Hence it is crucial to know how to please them, how to speak courteously and to negotiate fairly with them, for these beings also have a desire to survive and a need for sustenance, and their food is none other than the flesh, blood, and soul of human beings.

On this earth, then, there are only hunters and prey, eaters and eaten. The food chain must be fairly managed in a "generalized hunt" that joins humans to the Master of Plants and animals in one ineluctable cycle of life and death.

Nonhuman entities—plants, animals, and fish—that are hunted and eaten have both a physical body and an invisible double—in this case, Empu—that informs how much should be taken and how (i.e., hunted, fished, cleared). Interpretations of these broadly reciprocal human-nonhuman relations vary across Pala'wan subgroups and the scholars working among them. Nevertheless, the omnipresent Empu must be shown respect and appeased through offerings, lest violent and hungry entities devour Pala'wan flesh and souls. This cyclical relationship between hunters and prey, eaters and eaten, ultimately connects humans and nonhumans in both life and death (Revel 1998).

Benevolent and Malevolent Spaces

Beljan and other Pala'wan negotiate and experience these benevolent and malevolent invisible beings through ritual practices, dreams, and everyday chance encounters. The benevolent invisible deities—diwata or taw kewasa (the powerful ones)—reside in celestial spaces, mountaintops, and the flowering canopies of large honey-bearing trees such as Apitong (Macdonald 2007; Theriault 2017).[5] They are revered as guardians and caretakers of flora, fauna, and other objects central to Pala'wan beliefs and livelihood (e.g., honey, rice, or baskets) and are called upon during rituals to heal and bring clarity to uncertainty. They are often invoked with offerings during harvesting and healing ceremonies to convey gratitude and bring good fortunes.

These taw kewasa are described as invisible, humanlike figures (Macdonald 2007), often "people of the forest" (tawa't gebaq or taw't talun) who reside in and use older forests, groves, single trees, and the forested banks of rivers (Macdonald

2007; Theriault 2017). As Theriault (2017, 121) notes, they are "unambiguously human and thus prone to anger and jealousy," particularly if Pala'wan clear forest areas for swidden without consent or overharvest important species (e.g., rattan and honey). Such breaches, or even accidental encounters, can lead taw't talun to take up residence, impede certain livelihood activities, and inflict serious illness and death.

Several other dangerous subgroups of invisible forest peoples (or entities) exist. The Meliwanen are beautiful people who seduce and kill Pala'wan who trespass in or clear "taboo" (*lihien*) forest groves without permission, while the Mengeringen inhabit large rock outcrops or boulders covered in pandan (*Pandus spp.*) that are always off-limits. Those who breach their sanctum are cursed with leprosy (Macdonald 2007, 103). In Kamantian and Marenshewan, the most frequently mentioned variety of taw't talun was the *lenggam*, hostile demons who eat Pala'wan, make them seriously ill, or bring madness. The lenggam jealously guard lihien forests or occupy their own swidden fallow or free-standing balete trees (*Ficus spp.*) in cleared fields. When disturbed or violated, they become crazed and vengeful. They may roam with dogs and unleash them upon Pala'wan who violate their homes (Macdonald 2007, 103). The lenggam are also caretakers of venomous and biting creatures, such as centipedes and snakes, sent to punish humans who have transgressed (Novellino 2001, 83). Other malevolent entities include the *Gila'en* (the mad ones) and *tawa't lingeb* (invisible people of darkness): underworld beings who roam freely through older, darker forests. If their space is violated, an offering (*ungsud*) of a white chicken or equivalent (e.g., gongs, beaded necklaces, or first rice) must be made to ensure the offender's kurudwa does not go astray.[6]

The reproduction of Pala'wan beliefs emerges from a living spiritual landscape that nourishes the livelihoods of individuals, kin, and community in the uplands. The many invisible beings that lurk within Pala'wan forest landscapes influence everyday practices and, in many respects, nonstate interventions.

Rice as Personhood: The Life Force of Kurudwa

The construction of personhood informs Pala'wan relationships with the nonhuman world, particularly with certain animals (e.g., wild boars) and plants (e.g., rice), and further influences how they perceive swidden and other livelihood activities (Macdonald 2007; Theriault 2017). Nonhuman entities prefixed with *taw't* possess human consciousness and emotions such as madness and distress that allow them to interpret people's actions and respond accordingly (Macdonald 2007; Novellino 2015, 109). The Pala'wan ascribe the qualification *taw't* (*taw*, *tao*, a human, a person) to various superhuman or nonhuman agents who are addressed during rituals (e.g., taw kewasa), everyday encounters, and dreams (e.g., taw't talun) (Macdonald 2007). Becoming or existing as a (good) person is thus attributed to one's relationship with the disposition and vitality of nonhuman species, particularly swidden rice.

Swidden (uma) cultivation and dry rice offer the Pala'wan powerful means of connecting with the nonhuman world through various rituals and myths. The beljan mediate such nonhuman relations by apprehending guidance from spirit-beings about whether older taboo (lihien) forests should be cleared for swidden (or harvested for NTFPs). While dreams provide the vehicle through which such messages from the invisible spirit world are received, beljan and others elicit such encounters through the ritual offerings (ungsud) of *lutlut* to determine whether lihien forests can be cleared, cultivated, and harvested (Dressler 2009; Smith 2015, 2021; Theriault 2017). Lutlut consists of glutinous sticky rice (*maregket*) cooked in bamboo internodes, which, as a divine food, is offered to appease the diwata (often Empu et Parey, the Master of Rice) and ascertain certain socioecological outcomes (Smith 2015; Dressler et al. 2018).

Pala'wan reverence for rice is reflected in its mythic origins in human sacrifice and care—the very basis of producing food and becoming human (Macdonald 2007; Novellino 2011, 2015; Smith 2015, 2021). Across the highlands, elderly Pala'wan recount varying tales about the origins of rice, where the Master of Rice, Empu et Parey, instructs a Pala'wan family to sacrifice their youngest child or another family member as punishment for incest (*sumbang*). This significance of such an offering yields a unique, sentient rice crop, setting it apart from all other foods. Beljan often talk of a female figure —Manalib or Maraga— who was visited by Empu 't Parey in a dream. Empu promised a stable source of food beyond forest edibles and root crops, but she must sacrifice her only child and scatter its blood, bones, and body parts across the landscape. Three days later, rice grains sprouted from the remains of the child, with the blood producing glutinous black and red rice varieties (*maregket itum/pula*) and from the bones, glutinous white varieties (*maregket puti*). The other body parts grew into nonglutinous varieties (*marenggas*) in the swidden (Smith 2015, Bibal 2023). This mythical sacrifice imbues all rice with humanlike consciousness (kurudwa) and divinity (Macdonald 2007, 125, 279). Rice plants retain a humanlike essence; they can speak, and they can experience a range of emotions. Many Pala'wan farmers consider rice to be human (*taw*), soulful, and close to Empu.[7] Closer to Marenshewan, panglima Palinang describes the human character or consciousness of rice. In greater detail, he describes how rice must be cared for like a child:

> Empu't Parey is like a person to us. It's just that we cannot always see him. *Parey* is our story. We have native [heirloom] varieties, which are

like his children. He is not careless, that's his promise to us. In the past, there was no rice and uma clearings yet, we only ate mushrooms from the forest as well as sweet potato, cassava, and banana. . . . It was then that Empu't Parey took pity on us and said, "If you really want to have parey, then you have to kill your own child." So Empu 't Parey commanded a man in his dreams to do this. The man did not think twice, he got the *aratan* plant with long thorns, and when they arrived at his uma, he thrashed his child. After a week, many rice plants grew at the same spot. "You can disapprove of this," he said to his wife, "but the parey is already flowering and bountiful." But the husband had not informed his wife that the parey was their child. The wife just asked, "Where is our child?" "I don't know," the husband said. But she [the wife] asked, "Where did you get this?" He said, "I don't know . . . it just grew like that." So they harvested the rice [child]. Upon holding the harvested rice, he tried to cut it, and the rice said, "Ouch, you are hurting me." From then on, his wife knew that the rice was their child. Once the rice was cooked, the pain disappeared.

Kurudwa thus enlivens and animates not only rice but all living things (Macdonald 2007). The essence of kurudwa reflects a person's soul or spiritual components that continue to exist after the individual has died (126); it is distinct from *ginawa* (a life force or "breath of life," responsible for corporeal existence) and *nakem* (awareness, mindfulness), which do not survive a person's death (Iskander 2016, 4).[8] Kurudwa, which is found in various body parts, may leave the body during dreams and trance states. It can also be "dislodged" due to accidents or illness (Iskander 2016, 4) or leave the body when sleeping or consumed by taw't talun or other malevolent deities upon encountering them in the forest. Losing one's kurudwa in this way can cause illness and sometimes death. Several kurudwa (e.g., in the feet, hands, and ears) are known to be difficult or bad, while the one occupying the forehead is considered benevolent and likely to reach Empu after death.[9] Any kurudwa that transcends the human body reflects an agentive force that, enlivening plants and animals, exerts a strong presence in forests, fields, and fallows (Macdonald 2007, 125).[10]

It is central to the beljan's mediation of rituals to heal those suffering from illnesses inflicted by lenggam in everyday and dream encounters. Novellino's (2001, 86–87) vivid account of a beljan's diagnostic and curing practices explains how the healer treated an ill woman of the *Taw't Batu* subgroup in a remote Pala'wan territory. During a healing ceremony, the beljan used coconut oil (*lanaq*) to wet the powerful and fragrant cross-ceremonial plant *ruruku* (basil, *Ocimum sanctum*; see fig. 2.2) and a ritual object for dancing (*tarek, dung dung*) made of sila leaves (*Licuala spinosa*) frayed and bound at the end.[11] The coconut oil was placed on the beljan's feet, forehead, and whorl of the hair: the locations where kurudwa enters and subsumes the body. The scent and oil of ruruku were also rubbed upon and attached to the beljan to endow him with greater clairvoyance and enable his kurudwa to leave his body.

As gongs and drums played with increasing intensity (*basal*) and the customary *tarek* dance began, the beljan's kurudwa began its journey to find the entity responsible for the illness. The sick woman asked the beljan questions about her illness, including how she might appease the entities responsible. In a trancelike state, the beljan responded that her illness had been caused by an unintentional encounter with lenggam in the forest; the woman had brushed against the lenggam, causing the demon's hair to enter her body, where it festered and caused her to become ill. The beljan then offered instructions about the type of offering required by the deity. Fortunately, the lenggam had not stolen her kurudwa outright.[12]

These descriptions reveal the complex vicissitudes that beljan negotiate as they diagnose uncertainties, moral transgressions, and illnesses. Such customary practices are critical to the social reproduction of Pala'wan life and livelihoods in

FIGURE 2.2. A sprig of *ruruku* (basil, *Ocimum sanctum*), 2013. Photo: Dressler.

the uplands and contrast powerfully with nonstate actors' biopolitical logic and practices.

The Pala'wan Swidden Cycle

Making Uma

As in the past, the annual swidden cycle figures most prominently in the social relations, livelihoods, and lifeworld of Pala'wan living in the interior of the southern highlands, less so in the midlands and lowlands (Macdonald 2007; Dressler 2009; Theriault 2017; Smith 2018). Pala'wan swiddens are complex socioecological assemblages of dry rice varieties, root crops, NTFPs, and streams that host diverse (often scaleless) species. They entail a mixture of field, fallow, and forests that underpins highlanders' self-sufficiency and autonomy. Varied and often fragmented ritual practices, dreams, stories, observations, and reciprocity mediate the swidden cycle's complex human-nonhuman relations. Constellations offer signals of when to burn and plant in line with changing weather patterns and hunger, and benevolent and malevolent deities influence which forest areas can be cleared without retribution.

The careful cultivation and maintenance of the superhuman worlds of swidden across forest landscapes underpin the prudent harvesting of diverse species for food, healing, and ritual. This is the anathema of the dietary and conservation objectives promoted by the Adventist Frontier Missions (AFM), Conservation International (CI), and the Palawan Tropical Forest Protection Programme (PTFPP). The character, quality, and quantity of what Pala'wan harvest and cultivate informs the basis of their existence and nonstate actors' targeted reforms.

At its most basic level, the annual swidden cycle begins in December or January and ends in late September or October. The cycle broadly aligns with the beginning of the northeastern monsoon *amihan* (*emihan*) from October to May that brings dry winds, heat (*bulaq/g*), and sun needed for effective drying and burning of debris, and the southwestern monsoon *habagat* (*barat*) from June to October that brings warm, humid air and the heavier, persistent rains needed to nourish recently planted rice, root crops, and other saplings.

The cycle involves site selection (*ternaban*), repeated clearing and underbrushing of smaller trees (*ririk*), clearing of large and medium-sized trees (*penenbeng*), iterative burning of dead forest debris (*d/ruruk*), sequenced sowing or planting of seeds (*sasad*), repeated weeding (*mengilamun*), and several years of harvesting (*pangteb*) field and fallow (Revel 1990; Macdonald 2007; Cramb et al. 2009; Dressler et al. 2021). Uma farmers carefully observe weather, pests,

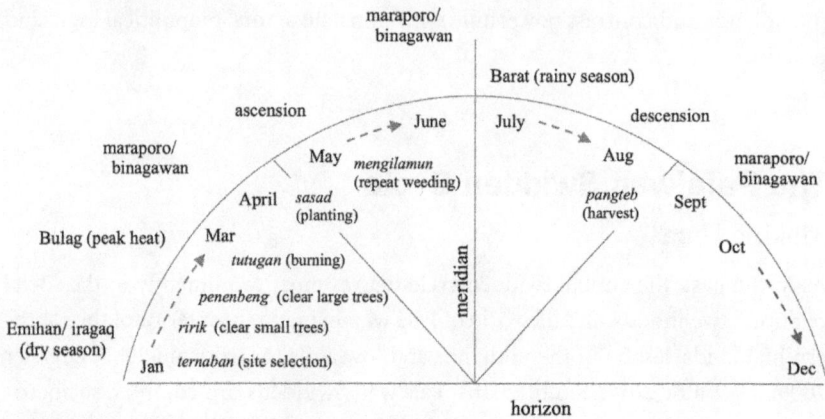

FIGURE 2.3. Pala'wan swidden cycle in line with seasons and constellations. Dressler field notes, 2023, adapted from Revel 1990.

and the condition of fields, while honoring the sentience of rice and other crops, and the myths and movements of two constellations. The Maroporo (the Seven Sisters, Pleiades) and Binagwagan (Balatik, Orion's belt) asterisms' rise and fall across the horizon over six to seven months marks *emihan* (January), the coming *bulaq* (late January-March) and *barat* (July) and underpins the basis of the myths (*tuturan*), events, and practices for field preparation (see fig. 2.3).

A successful swidden cycle depends on several interrelated sociocultural, economic, and environmental factors: family size, needs, labor availability, soil quality, the age and complexity of the forest/fallow, field size, the quality of the clearing, the burn, the amount, and type of rice planted, field weeding, pests, guarding, weather conditions, population dynamics, and external interventions, all play a role. Although they are carefully managed to overcome social and environmental uncertainties, swidden rice yields and other crop harvests are declining in southern Palawan, partly due to nonstate designs.

Site Selection, Clearing, and Burning

Uma site selection (*ternaban*) begins in late December or January and involves different ritual "divination" practices that aim to ascertain whether the rice cropping will be successful (Macdonald 2007, 114). The *luslusan* ritual, for example, involves rubbing and placing stems cut from the *buldung* (*Amarantha donax cannaeformis*) plant eight times in the field. If the cuttings grow several inches more than the original length, it is a sign of fertile soils and, possibly, a bountiful rice yield. Crucially, ternaban is informed by the arrival of emihan and changes in the constellations. Pala'wan farmers will wait for the appearance of Maroporo emerging from above the horizon as a signal to select the site of the uma, and then, as Binawagan appears alongside, they know that the dry season is approaching. A range of other factors from household needs, kinship principles, residence patterns, forest cover type, and the mood of deities residing there further influence uma size and location.

Able-bodied Pala'wan men typically select sites that range between one quarter and one hectare in secondary growth forests (*bunglay*), which are considered less laborious and legally permissible to clear and burn. Previous generations were less concerned about the perceived illegality and labor investments required to clear and burn swidden plots in older growth forests (*gebaq*). Indeed, Pala'wan elders knew that older forests lock in thick biomass that, when mixed with nutrient-rich ash, yields bountiful harvests of rice, root crops, and other cereals, with favorable climatic conditions (Dressler 2010; see table 2.1). In areas where land is abundant, farmers tend to avoid shorter two-to-three-year fallows because of the associated decline in soil fertility, significant weeds, and pests that result in low yields (Macdonald 2007; Dressler et al. 2019).

Most farmers prefer to locate their fields near their families' or extended kin's fields in accordance with matrilocal residence patterns, labor obligations, social ties across genders and generations, and forest characteristics (Brown 1991, 196–200). Older farmers who are too physically weak to work and younger households with children to care for, may clear fields in young fallows or secondary growth forests near their homes due to ease of access. They may also be assisted by family members in preparing and managing fields in more remote, mature forests. More self-sufficient and able middle-aged households may clear fields in mature secondary growth forests a fair distance from their homes, where land is more abundant. Such fields are typically located on moderate-to-steep slopes in the interior. The edges of a cleared swidden plot are established using the landscape's natural features—rivers, larger boulders, and well-positioned trees.[13] A family's swidden cycle reflects its socioecological signature across the forest landscape (see fig. 2.4).

Certain portions of forest are lihien (forbidden), the abode of malevolent or capricious invisible nonhuman beings (Macdonald 1997; Smith 2015; Theriault 2017). Farmers who breech or accidentally disrupt the Mengeringen or other taw't talun (invisible people of the forest/trees) residing in large balete trees, bushes, or cavernous rocks in or near an uma violate the malevolent entity's abode. They may become seriously ill from the encounter, experiencing high fever, sweating, chills, pain, and lethargy. In serious cases, the spirit of the taw't talun may displace the farmer's kurudwa, leaving them incapacitated or near death. Most farmers thus avoid clearing lihien forests (and/or similar areas) occupied by dangerous forest deities, leaving groves uncleared within a mosaic of differently aged forest fallows (Brown 1991; Smith 2015; Dressler et al. 2017). As the elder Marwang Bilod, from the uplands near Mount Mantalingahan, noted:

> If you are in the forest and you get sick, it will be from the mengeringen; they are saytan [evil beings] that cause the sickness here in the mountains. If you do ririk in the abode of a mengeringen and you did not notice them, the unseen being will get angry and inflict an ailment upon you. You will get sick, and they will be angry because you disturbed and disrupted their forest area. . . . You cannot see them and when you cut down their forest home, you offend them, and you will get sick.

The time for clearing (*ririk*) comes in January and February, as Maroporo and Binawagan gradually ascend to the northeast.[14] After establishing the swidden fields' spatial dimensions, Pala'wan men begin by clearing small and medium-sized trees, brush, and vegetation (ririk), working from the exterior to the interior with machetes. They then use an axe (*kapa, wasay*) to cut larger hardwoods, whose foliage shades the rice and whose roots make planting difficult (Revel, Xhauflair, and Colili 2017). After constructing and mounting an elevated stilt, they cut above the trunk's wide flute.

FIGURE 2.4. A cleared and burned swidden field in Marenshewan (800–900 m asl), 2013. Photo: Dressler.

When both constellations rise just before daybreak (heliacal rising), it is a clear sign of the "beginning of heat" in March (Revel 1990). As the northeasterly trade winds (*iragaq*) of the emihan monsoon set in and bring the peak of the dry season (bulaq), Pala'wan may call upon the Master of the Northeasterly Trade Winds, Bulalakaw, for divine intervention—robust dry winds that intensify the first burning of the fallen debris from the initial uma clearing. The elder Gantilo Libasa, from the interior highlands, recounted what he learned from his father about bulalakaw: "When we set a fire, we wait until the winds (*deres*) are strong so that the fire will continue to burn and call out: 'You, bulalakaw, here is my uma, set this on fire,' that's what we do. When it's noon, we fire the uma and call bulalakaw. We just call and call."

Most burning commences in March, the hottest month. The bulaq is a much drier, hotter period with only intermittent rains that signal when the uma should be burned and planted (Revel 1998). The best time to burn is when debris, leaf litter, and soils are sufficiently dry and brittle. Both small and large trees are cleared and burned to render ash, with the burn's effectiveness determined by the type, amount, density, and dryness of debris (e.g., smaller logs may be placed around larger stumps to ensure hot, efficient burns in older forests, which dry slower, requiring hotter burns; see Dressler et al. 2021). Each farmer carefully considers slope, field position, wind direction, time of day, and surrounding forest type in deciding how to burn. Most construct a firebreak (*gahit*) two to three meters around their field and use a firebrand (*ta'ta*) to burn from the edges toward the interior in the general direction of the wind. As the flames move inward, they engulf large amounts of strategically placed debris. After the main burning, any poorly burned or unburned debris is repiled and burned again at the center or edges of the field (*duruk*).

Uma Planting and Maintenance

In April, if both Maroporo and Binawagan are in full constellation and moving toward the horizon again, farmers know to begin sowing rice seeds (*sasad*) and other root crops in their uma before the southwestern monsoon. Burning too early or late can lead to poor yields and significant food insecurity in the following year. Burning and planting too far ahead of the barat rains can lead to excessively dry soils, with ash blowing away and crops dying. Planting at the onset of the monsoonal rains can lead to poor burns, limited ash deposits, less fertile soils, and significant weeds (Smith 2015). Ideally, then, rice is planted in the field several weeks after burning and well before the heavier rains of barat.[15]

Just before the rice planting, Pala'wan will build a *pinadungan* structure in the middle of the field, signifying a divine "seat" for Empu et Parey (Macdonald

2007, 114). A small stand or table held by four posts holds a *tabig* basket full of rice seeds that will soon be used for planting. Ruruku is tied, and *Tanglad* (lemongrass) is planted, at the base of the posts to ward off malevolent deities and to ensure the vigorous growth of the rice. After Empu et Parey and Empu et Lugta (Master of the Land) are called upon, rice seeds are planted around the pinadungan and in eight holes at the field's edge toward the setting sun—an offering to the dead, or *lihej* (Macdonald 2007, 115). The latter planting is reserved for the ancestors and held separately.

Blessed by Empu et Parey, the rice seeds placed upon the pinadungan are treated with great care for their kurudwa must be protected. As one panglima noted: "Empu't Parey said, 'There must be uma, don't take this lightly, the rice there is my child!' There should be a pinadungan in the middle of the uma, where you can place the parey for when you do *sasad* (plant). This way the kurudwa of the rice will be taken care of—it will not be driven away."

The divine rice seeds are then carefully planted in the recently burned field. Men use a dibble stick, or *tugda*, to puncture the soil, while women follow close behind, placing rice grains in the holes, or *sasadan* (see fig. 2.5), though these gender roles are often reversed. Traditionally, up to seventy varieties of rice were exchanged and planted (Revel 1990, 339); today, however, only five to ten varieties (heirloom, high-yielding, or both) are planted in any given field. Different rice

FIGURE 2.5.   Pala'wan men and women planting rice in a partitioned, burned swidden field, 2018. Photo: Dressler.

varieties are almost always planted in segregated areas marked with sticks (*bental*) for easier identification during harvest (e.g., glutinous varieties, or *maregket*, are kept separate from nonglutinous varieties, or *marungras*).

During the rice planting, specific crops are often planted where larger tree trunks, long logs, and piles of branches were burned out to take advantage of the rich, moist soils with deep ash deposits. These include various mixtures of purple yam (*ubi, Dioscorea alata*), sweet potato (*sanglay, Ipomoea batatas*), squash (*labuq, Cucurbita maxima*), taro (*tales; Colocasia esculenta*), pineapple (*Ananas comosus*), and ginger (*Zingiber officinale*). A farmer may also plant banana suckers (*punti*), various taro and cassava shoots (*sanglay kayu, Manihot esculenta*), millet (*aturay, Setaria italica*), sorghum (*batad, Andropogon sorghum*), sugarcane, and other taller crops at the field's edge to avoid shading or crowding the rice. Juvenile cassava can also be intercropped with rice (Novellino 2011; see table 2.1).

Preparing and maintaining the entire uma is a familial and gendered affair, with women and sometimes men weeding at the beginning, middle, and end of the swidden cycle. Weeding is a laborious, time-consuming task that is only made easier if children assist their parents. Larger fields cleared in less mature secondary growth forests produce significant weeds after clearing and burning. As the rice and other crops mature, the family and their kin may reside in a

TABLE 2.1   Select list of primary Pala'wan root crops, legumes, and cereals

ENGLISH VERNACULAR	PALA'WAN	TAGALOG	SCIENTIFIC
rice (various varieties)	*parey*	*palay*	*Oryza sativa*
sweet potato	*sangley*	*kamoteng baging*	*Ipomoea batatas*
cassava	*sanglay kayu*	*kamonteng kahoy*	*Manihot esculenta*
purple yam	*ubi*	*ube*	*Dioscorea alata*
taro	*tales, bangkuka*	*gab, Palawan gabi*	*Colocasia esculenta*
sorghum	*batad*	*batad*	*Andropogon sorghum*
millet	*aturey*	*atoray*	*Setaria italica*
corn	*mais*	*mais*	*Zea mays*
pigeon pea	*kedyes*	*kadyos*	*Cajanus cajan*
banana	*punti*	*saging*	*Musa spp.*
squash	*labuq*	*kalabasa*	*Cucurbita maxima*
peanuts	*belatung*	*mani*	*Arachis hypogaea*
pineapple	*peranggiq*	*piña*	*Ananas comosus*
ginger	*luya*	*luya*	*Zingiber officinales*
tobacco	*sigup*	*tabako*	*Nicotiana latissima mill.*
pepper	*rampa*	*paminta*	*Piper nigrum linn.*
maize	*mais*	*mais*	*Zea mays*

Sources: Dressler field notes; Macdonald, 2007; Dressler 2009; Smith 2015; Dressler et al. 2016.

nearby temporary thatch hut (*kubo kubo*) to guard against pests like rats, rice ear bugs (*tayangaw, Leptocorisa oratoria*), and marauding wild pigs (*biyek, Sus ahoenobarbus*). Any wild pigs and monkeys frequenting the field in search of succulent sweet potato and morsels of fruit are scared away, trapped, and/or killed for meat (sometimes using homemade "pig bombs"). The smaller, raised granaries (*lagkaw*) are used to store rice near the fields and homes, sheltering it from rats, monkeys, and birds and, importantly, protecting its kurudwa. In the mid-to-lower elevations, Pala'wan households with extra income occasionally purchase fertilizers and pesticides to ensure good yields. However, in the uplands, most farmers employ more creative strategies to repel pests (e.g., burning rubber sandals and piling wet debris in the middle of the field; see Dressler et al. 2021).[16]

Uma Harvesting

The uma rice harvest (*pangteb*) is staggered between July and October and some-times marked by further divinification of the *pinadungan* and *lagkaw* (rice gra-nary).[17] Before the harvest begins, a farmer will place a portion of a louse in beeswax at the bottom of a tabig basket wrapped in cloth before four male and four female rice heads are added. The basket and its divine contents are eventu-ally tied to the roof of the lagkaw to care for the kurudwa of the rice and show respect to Empu et Parey (Macdonald 2007, 114–15). As one panglima noted: "But to be sure, the rice must also be placed in the lagkaw so it will be safe. . . . The kurudwa won't leave, it is nurtured there. The rice will grow better." The rice stalks from the previous ritual planting around the pinadungan are then tied into bundles and fastened to the dibble stick, which is then fastened and erected in the center of the pinadungan. The harvest begins after this is complete (Macdonald 2007, 114–15).

The harvesting (and planting) of rice is a reciprocal affair between human and nonhuman beings in upland sitios; Pala'wan give thanks by sharing por-tions of the rice harvest with others and spirit beings, sharing rice grain as seed and as offerings, and providing labor when others begin the swidden cycle. The rice harvest is typically staggered: the first cropping (*inunahan, inuna'an*) of shorter, faster maturing varieties is done after two to three months from July to August, while the main harvest (*inurwayan*) of longer rice varieties (includ-ing maregket) occurs in September or October to avoid the barat's heavier rains (Novellino 2011; Smith 2015; see also Fox 1954 for Tagbanua). Most other crops planted around March are harvested from July; any intercropped cassava or other species is harvested somewhat later, or as needed during times of hunger (tag *urap*), when rice stores are lowest in July and August. No matter the vari-ety, rice is harvested with caution and care so as not to offend Empu et Parey (Macdonald 2007).

Swidden rice remains the most desired and significant staple crop for caloric intake and customary and ritual practices. Almost all farmers believe that cooked uma rice (*linamuq*) alone can satiate hunger and provide sufficient strength for a day's labor. The main postharvest ritual includes an offering and consumption of lutlut cooked over embers with glutinous maregket, coconut milk (or water), and sometimes meats in bamboo internodes (Smith 2015). The first grains from the main harvest and other important household goods (e.g., tobacco and sardines) are presented as offerings to Empu et Parey to ensure a successful harvest the following year.

Shorter Fallows, Declining Harvests?

Decades ago, Pala'wan farmers only returned to a field after leaving it fallow for eight years or more, making a rice yield of up to 800 kg/ha relatively common. Today, shorter fallows of four to five years are more common and produce significantly lower rice yields (McDermott 2000; Dressler 2009). For example, in 2011, the average swidden rice yield per hectare was a meager 259 kg in Marenshewan and 135 kg in Kamantian (author's observations). Similarly, Smith (2015, 84) reported a low average yield of 323 kg of rice per hectare in Inogbong in 2010. In central Palawan, Tagbanua and Batak swidden rice yields also showed an average decline from 425 kg of rice per hectare in 1980 to 159 kg in 2009 (Dressler 2014).

Farmers will compensate for lower rice harvests by planting other faster-growing and maturing starchy staples, such as cassava, corn (*mais, Zea mays*), peanuts (*mani, Arachis hypogaea*), and tobacco (*sigup, Nicotiana tabacum*) in sections of their uma (see table 2.1). These crops are harvested two to three months later and may be sold in lowland markets (e.g., cassava and sweet potato) or used for family consumption when core staples run low during the year, depending on the rice surplus or cassava stores.[18] In September and October, the middle and outer sections of the field may be recleared—a practice known as *pengengawat* (or *dab-dab* in other upland dialects)—to plant root crops and legumes, including sweet potato, squash, beans, and yams.

These broader declines in swidden rice yields across highland areas are attributed to intersecting socioecological and biopolitical governance pressures (Dressler 2009).[19] Nevertheless, swidden production remains central to the social reproduction of Pala'wan uplanders and is, therefore, still targeted by nonstate reforms seeking to sedentarize and proselytize Indigenous uplanders.[20]

Fallow Life and Nontimber Forest Products

Contrary to nonstate (and state) actors' widespread misconceptions, swidden fields are not simply abandoned as idle and degraded forest lands (Dressler et al. 2018).

Rather, swiddens are actively maintained over multiple decades as lively fallows that coproduce agroecologically diverse forest landscapes. The short- and longer-term use of a swidden fallow depends on what was cropped and the overall length of the fallow period. Although many fields in Kamantian and elsewhere are in reduced fallow and harvested annually or even biannually, others remain in a fallow state for seven years or longer, depending on the field's location. After most rice and root crops are harvested, the various fruit trees, cassava, and other crops will regrow, surrounded by forest vegetation. After several years, large cassava "trees," fruit-bearing trees (e.g., mango, kalamansi, and jackfruit), and other crops like sugarcane and pineapple may merge with other tree species into a young second-growth forest (bunglay).

The initial field also regrows as a diverse polycropped forest fallow that can serve as a food store in times of need. Pala'wan families across highland areas use these fallows as spatially dispersed and diverse agroecological "caloric" repositories (Brookfield and Padoch 1994). Although claims to fallows decline as forest regrows, most families retain fallow tenure in secondary growth forests. They return to harvest key starchy staples like cassava and other root crops that grow over longer periods (but become bitter) during food shortages resulting from poor harvests, inclement weather, family emergencies, or when retreating from conflict situations (Dressler 2009; Smith 2015; Dressler et al. 2016; Scott 2009 on "escape crops").[21] Farmers also cut, crop, and transplant stalks of taro (usually from the wetter soils), banana, and cassava (usually from drier soils) between fallows and in new settlements, producing a "mobile kitchen" in times of scarcity. Finally, Pala'wan farmers may clear certain sections of a fallow for new swiddens or draw on the fallowed forest landscape for NTFPs.

More than Just Stalks and Roots

Swidden fallows are more than stalks and roots; Pala'wan women and men harvest an abundance of nutritious and useful items in the depths of older fallows, typically neglected by nonstate actors. Mature fallows and secondary forests (banglay) harbor complex and diverse forest socioecologies used for both food and shelter. Older and younger women harvest bamboo and bamboo shoots (rebuk, Bambusa), anibong (Anibung, Arecaceae), vines (gahid), and different types of honey (deges). From this bounty, they will craft various items essential for custom and livelihood, including the lidless basket, the tabig; variously sized Tingkep; and the winnow, the nigu, among others.

In older forests (gebaq) that blend into and emerge from younger secondary growth, younger, middle-aged Pala'wan men harvest both rattan (uway, iyantok, Calamus spp.) and almaciga resin (bagtik, Agathis philippinensis), despite the physically demanding work and generally low return.[22] Both harvests provide

Pala'wan families across the highlands with an important—albeit precarious—source of forest-based income.[23] Alone or in hunting parties, these men will also hunt diverse fauna in fallow mosaics, most importantly wild boar (*biyek*), squirrel (*buyayak*), different avifauna, and various bats (*ememkung*; e.g., *Ptenochirus jagorii* Peters).

Meanwhile, Pala'wan boys hunt smaller birds, and girls forage for insect larva, freshwater eels, crabs, shrimps, snails, and smaller fish (*Gobiidae*) in and along the well-shaded streams and rivers that cut through lush, moist forest fallows. Among the larger trees, young men use blowpipes (*sapukan*) to hunt songbirds and smaller mammals (e.g., squirrels), which are often placed, bloodied from darts or arrows, in the tingkep to be carried home (Revel, Xhauflair, and Colili 2017). This gendered work of NTFP harvesting, tingkep weaving, and swiddening confounds nonstate actors' attempts to narrow or even erase the Pala'wan socioecological repertoire for their respective biopolitical agendas.

Tingkep Origins and Forest Assemblages

No single story narrates the tingkep's origins. Rather, the Pala'wan elders I spoke to offered various narratives that converge around the basket's shared characteristics, origins, meanings, and materiality in ritual and livelihood. The tingkep's origins center on the "World of the Creator" (Empu Banar), the capricious taw't gebaq/talun (invisible people of older forest/forests), spiritual deities (e.g., diwata, Linamin) and diverse forest species. Pala'wan accounts of tingkep origins not only relate to Empu but also recount how taw't gebaq mediate and facilitate tingkep's powers across the corporeal and spiritual worlds—worlds beyond lowland markets.

In hushed tones, elderly Pala'wan spoke of a lone, crazed, invisible entity with humanlike characteristics who wanders, day and night, through the forests to produce the basket based on what is seen and sensed. This malevolent entity resides in older forests and shape-shifts from a humanlike form into that of an anthropophagous beast with a long tail and sharp fangs. Although younger Pala'wan seldom recall his name, older beljan and weavers refer to this shapeshifter as *Gila* (crazy) or *Gila'en* in reference to his crazed, twisted state (and potentially the fate of the weaver). Elderly Pala'wan regard gila'en with deep fear and respect. A well-known female weaver, Pelima Apol, described how the gila'en's madness leads him to wander through the forest and receive signs from Empuq about which plant materials to use in weaving the baskets—everything from rattan (*iyantok*, *malahog*, *arurung*) and rattan vines (*megkelawit*) to bamboo (*binsag*) and softer wood (see fig. 2.6). Others suggest that Empu informs gila'en of important weaving and design patterns that might be drawn from complex patterns found in

FIGURE 2.6. Pala'wan elders showing their craft, 2013. Photo: Dressler.

forests. As gila'en gazes into the forest, he creates weaves, patterns, and colors for baskets from the understory assemblages.

The gila'en was born insane and creates baskets from his madness; he is possessed by evil spirits (*pangpaning*) and has lost himself in madness (*dar/tdalit*). If a basket is left incomplete or improperly woven, the gila'en will seize the weaver's kurudwa and inflict crazed madness (gila gila) upon them. Some elderly Pala'wan suggest that gila'en presents them with unique designs in their dreams or writes designs into the forest landscape for them to follow. However, caution is advised. As elder Hamma Tiblic warned during a lengthy conversation, "If you follow the designs in your dreams, then you must do it well, or you will become possessed by gila and go mad." Similarly, elder Donisia Masad points out that "if you cannot finish making the Tingkep, even if you do it seriously, you will go insane . . . so, whatever you do, finish it right away so you don't get caught by the gila'en." Other elderly weavers, such as Marsa Apel, stressed that "if you are able to make seven tingkep then you will be OK, the madness will leave you." The correct and complete weaving of seven baskets or more—reflecting the celestial layers of the Pala'wan universe—brings good fortune from Empuq Nagsalad, "the Weaver," who "with seven golden threads . . . wove the world" (Revel 2011, 50). The good fortune from Empu, based in the divinity of completion and creation, dispels such fits of insanity and ensures that weaving practices and patterns are respected and sustained through generations.

Learning how to craft tingkep is a serious social affair that is not taken lightly; the gila'en strikes fear and desire into potential master weavers, both young and old, woman and man. Questioning or refusing the gila'en or looking him in the eye can cause one to fall sick, be transformed into a beast, or have one's soul claimed. One can even be eaten alive. As elder Dionysia Itom emphasized, "The gila said this is the only tingkep you must follow: don't ever laugh while doing it, make it one by one, two by two until you have learned, and then you can pass it along to the present generation."

Constructing Patterns of Nature

The name *tingkep* is derived from *tinakep* (covered) or *pinegtakep* (used to cover), in reference to its lid (NTFP-EP, 2008, 20). The tingkep is forged through complex sociomaterial assemblages of forest materials, human and nonhuman relations. Weavers scan the forest, seeking the bamboo and rattan species used to make the basket's body, skin, and exterior braces. Supple but strong bamboo species, such as *busneg* and *binsag*, are stripped and shaved (*meglegis*), and the soft pith (*embukan*) is removed. The strips are then woven together in a single (*sembatu-sembatu*) or double bicolored overlapping weave (*dwa-dwa*) to form the basket's skin (*bilug et tingkep*; *busneg* is blackened with resin [*saheng* or *almaciga*] soot, smoke, or charcoal). Elders continue to weave until they reach the desired height (23).

The basket's four "legs" (*tiked*)—sturdy braces that run vertically to the basket's mouth—are made of robust rattan, such as *arurung* and *lebsikan*. The internal set of braces (*patideg*) runs parallel to the *tiked* to create a cross-base (*sulambew*) of palm (*betbat*) or rattan (NTFP-EP, 2008, 23). An upper brace (*sabet*) runs horizontally around the basket's upper section. The stitching (*pengma*), made from strong, fibrous *nito* vine, fastens the woven skin to the braces and the base and rim (*semalang*), to which the "lid" (or *takep*) attaches. The upper semalang is made from softer wood (*enapung*), bamboo (*bikal* and *binsag*) (NTFP-EP 2008, 1-6; Dressler 2019), and harder rattan (*parasan*) at the base. The frames create a strong structure that is well suited for carrying forest items (e.g., birds, bats, and herbs) and powerful *mutja*, and for ritual practices (see fig. 2.7, table 2.2, and chap. 4).

A Living Skin?

The tingkep's skin is alive. During an interview, master weaver Pelima Apol described how gila'en and the ancestors forge weaving patterns from unique forest flora and fauna: "He taught us the different patterns of the basket. He taught us *Mata Puney, Mata Uled, Linindog and Patyug*. He told us to follow these patterns. From his younger years, he already knew. . . . When we were younger, the Gila taught us in our dreams how to weave. The elders taught us, too."

FIGURE 2.7. Tingkep structure and materials. Dressler photo, field notes; NTFP-EP 2013.

TABLE 2.2 Tingkep structure and materials in figure 2.7

ELEMENT	FOREST MATERIAL
(1) *Semalang* (rim and base)	Bamboo (*busneg* and *binsag*), rattan (*parasan*), and softer wood (*bonut-bonut*)
(2) *Pengma* (stitching)	*Nito* vine
(3) *Sabet* (body brace, horizontal)	Rattan (*yantok, arurog*)
(4) *Tiked* ("legs," body brace, vertical)	Rattan (*yantok, arurog*)
(5) *Bilug 't Tinkep* (body/skin of tingkep)	Bamboo (*busneg* and *binsag*), rattan (*buldung, busnig*)

Pelima continued listing the most frequently woven patterns of moving natures, such as the mata puney (fig. 2.8), bangbabang, baltalbig, mata uled, and duen emagas. She described the "Mata Puney, 'the eye of the puney bird (dove)' . . . it is from the gila'en who looked into the trees and saw a puney bird—a bird with shiny green and blue feathers with a curious eye. As the bird always looked the gila'en in the eye, he saw a cross in the middle of it and said, 'Weave this.'"

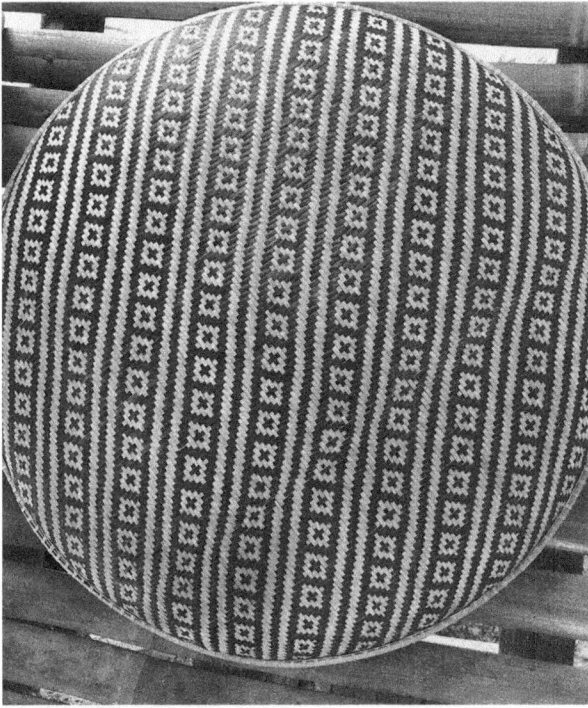

FIGURE 2.8. The Mata Puney weave, 2013. Photo: Dressler.

Another weave pattern, mata uled, resembles the eye of the groundworm. Pelima Apol revealed that it originated from "ancestors who would watch the worms eat their cassava and its leaves and copy the worm's eyes as it ate their root crops." She described another design rooted in fluid forest imagery (see fig 2.9): "The bangbabang is the wings of the butterfly. We got this one from our ancestors who copied the big wings of the bangbabang butterfly. They saw the butterfly sip the nectar from the flowers in the forest."

She explained that the baltabig is the oldest and most important source: "The baltabig is the color of a [wild] piglet's skin. It came from when my ancestors saw the color of piglet skin in the forest . . . they were then able to see all the other types of weaves in the forest. We copied the pattern of the skin of the wild pig." Other designs also stem from forest fauna and flora, including the deun emagas design derived from the leaves of the *emagas* palm tree (see fig. 2.10).

Similarly, the *law law* palm's black thorns and white leaves are reproduced in the tingkep's characteristic black-and-white motif, the linewlew. The tingkep's character thus embodies the vital confluence of the shifting human and nonhuman worlds of upland forest landscapes. In this way, it bridges Pala'wan lifeways

FIGURE 2.9. Bangbabang (or Binengbavang) weave, 2013. Photo: Dressler.

FIGURE 2.10. Deun Emagas weave, 2013. Photo: Dressler.

and forest natures. Its sociomaterial vitality extends further to the Kundu et Tingkep ritual practice—a ritual that powerfully reinforces Pala'wan social reproduction across the uplands at odds with evangelical ideals (Dressler 2019).

The Kundu et Tingkep Ritual

The tingkep is central to certain Pala'wan ritual practices. Organized by the beljan, the Kundu et Tingkep is a "focal ritual" that unites family, kin, and others together in the largest room of a family's house or larger meetinghouse (*kalang banwa*). The beljan uses a particular type and quality of tingkep to call upon the female forest spirit, *Linamen*, to provide answers to questions of familial or community concern (e.g., health, marriage, or weather). Before the ritual begins, as a full moon shines, the beljan dresses a small tingkep to resemble a person—for instance, as a figurine or doll. This tingkep is placed upside-down, with either side of the basket being fastened to sticks and a strong rope, which the beljan and others work during the ritual. The basket, draped with cloth (*kamut*), serves as the figurine's body; sticks are inserted to represent the arms, and a ball of material is placed on top to represent the head (Revel-Macdonald 1977). The sweet-smelling herb *ruruku* (basil, *Ocimum sanctum*) is either affixed to the basket or placed within it.

The figurine dances as the beljan chants prayers, calling upon the beautiful and dangerous female forest spirit, linamen et kunduq. Upon hearing the beljan's invocation, linamen is drawn to the sweet fragrance of ruruku and moves into the figurine. Linamen's power penetrates the tingkep and animates the figurine, sometimes accompanied by the burning of *parinaq* incense (*Kingiodendron alternifolium*) to attract the spirit (Revel-Macdonald 1977). Only smaller, well-worn tingkep—the kind souvenir shop owners call "antiques"—can be used for the ritual, since their smaller size and the smell of infused blood from forest game contains linamen's anger, movement, and escape. She does not recognize foreign containers, such as plastic buckets. The larger, lidless tabig basket is also inappropriate for the ritual, as it would allow linamen to become too powerful, thrust about, and escape to inflict harm on those participating.

Once linamen is drawn into the tingkep, the beljan brushes ruruku on the figurine's head and body as a blessing. Soon, the sounds of the *kudyapi* (or *kudlung*, the two-stringed lute) and *pagang* (bamboo zither) fill the house. Now, the kundu et tingkep unfolds in earnest as the beljan asks the figurine to dance the lively kunduq; the figurine sways in the air, vigorously dancing left and right, only touching the floor in time with the ritual tune, or *pagkulintangan* (Revel-Macdonald 1977). The kudyapi must be played continuously during the performance, creating an inseparable vital force within the basket, spirit, and music. Linamen (and the basket) is unlikely to dance without the kudyapi and directional chanting.

As this transpires, the Pala'wan attending the kundu et tingkep begin asking various questions about livelihood, health, or love. They may ask about the likelihood of wild game being caught in the forest, when inclement weather may abate, the general locations of plants and herbs required to cure a sickness, or whether the ill among them will die. The spirit moves the basket, tipping left for no or right for yes (Revel-Macdonald 1977). The ruruku herb that induced linamen into the basket also helps her detect who is sick during the ritual and heal them when the herb is later applied to their skin. As the male elder Persing Balabag narrated,

> The linamen of the forest will smell the ruruku flower and goes inside the basket. The limanen smells the sweet ruruku. . . . We cloth the basket, put a head on it, and then tie it like this, but if it is big we cannot hold and control it—if you loosen it, then you will become gila'en. We call and request for someone to be cured, so we must not be lying, otherwise it won't work. If I am true to her, and request to be cured, then our God will do it.
>
> The linamen dances to the kudyapi that we play as we call out a question about the sickness of someone, maybe a child. Sometimes she tells us who is sick amongst us. Only the beljan can see her. . . . Once the Tingkep falls over on its side, it means that linamen has left the basket to return to the forest. Only one tingkep can be used.

The kundu et tingkep is also performed to seek answers to any number of life questions unrelated to illness and healing—for example, who among the young Pala'wan attending will fall in love and marry? Who is the lucky one? (Revel-Macdonald 1977).

Mutja, Tingkep, and the Spirit Realm

Across the highlands of Palawan Island, Indigenous men and women will often keep *mutja*—esoteric material objects (Fox 1954, 176)—as powerful amulets (talismans or charms) that give the owner supernatural powers. These may be gems, curious stones, dried bees, the back hair of a wild pig, a pig's tooth or jaw, bird feathers, particular plants, or other objects. Often used interchangeably with the term *anting-anting* (more often used by Christian lowlanders; see chap. 3), mutja bestow Pala'wan users with extraordinary strength and success during the hunt (harvest, or yield), the ability to cure disease, or inflict illness upon others or—accidentally—oneself. Mutja are seldom used for both pernicious and benevolent purposes.

As Fox (1954, 177) notes for the Tagbanua, the mutja (also spelled *mutya*) "have varying degrees of intrinsic power and they are manipulated to control situations fraught with uncertainty"; their power is defined by "the social context in which they are employed and to the role and status of the user" (177). For Pala'wan, the esoteric value of mutja derives from its deep ritual and spiritual powers, both in mystique and application (e.g., handling and applying it during harvests may increase chances of success). Mutja may be used individually or in combination, and their potency is amplified by beeswax (*limbutan*) and *parina* incense.

Crucially, mutja are embedded within and embody the spiritual and supernatural properties of their provenance and the person who keeps and uses them. Although anyone can use mutja, Beljan draw on mutja to ward off or cure sickness attributed to malevolent invisible beings in forests, near rocks, and along streams. They may inflict illness upon someone (*sabat/sablaw*) or be used to influence people's social behavior to achieve a desired outcome during a discussion, debate, or broader arbitration (Fox 1954). As customary objects, they offer diagnostic powers and social meaning that contrast sharply with the reductionistic interventions of nonstate actors in the uplands.[24]

Tingkep et Mutja

Mutja also make the tingkep their home. The Pala'wan draw on the powers of mutja in their spiritual and everyday uses of tingkep to ensure that they successfully derive a livelihood from upland forests, particularly during the hunt. Pala'wan hunters often carry mutja from parts of the animal being hunted in smaller tingkep sashed around their necks and shoulders, believing it will attract the animal or strengthen the hunter during the chase and kill. Such mutja are made from commonly hunted fauna, such as the bristles of wild pigs or birds' bright feathers.

Across multiple conversations with elderly Pala'wan, I learned more about the powers of mutja. Pilmaya, a female elder, described how she uses special bird feathers to draw other birds when hunting them: "We used the mutja of bird feathers to catch a bird. . . . With the mutja the birds will come closer to you. We then kill one bird, and let it lie. And when another comes closer, alighting on a tree nearby, we try to kill it too."

While tingkep can hold a variety of personal possessions, Pala'wan elders and beljan use them to carry mutja with supernatural properties to ward off evil spirits. Elder Dorito Quinto shared a similar story, noting that "if there's mutja inside, it can ward off lenggam and help rid of sickness. If inside [the basket], no one can touch the tingkep. If the mutja is given to you, it cannot be used for evil purposes. You must forgive your enemies seven times." The number seven possibly reflects the layers of the universe transcending towards Empu's celestial abode.

Mutja taking the form of precious stones, such as gems, are also placed inside the tingkep, imbuing the owner with the power to ward off danger and attract good fortune. Elder Persing Balabag offered more insights:

> If you have the gem inside and you have an enemy, you will not be harmed or damaged. When they place the mutja in the tingkep, because we didn't have bags back then, they will utter a chant that will draw good and ward off bad. Other *anting-anting* were so strong and effective that it made things smoke in the basket. So, we did not bring it home. We would leave it in the forest because our children and spouse might be affected.
>
> The mutja is what we call *sukang-sukang* [causing an unwanted effect, get rid of] which is very effective. If we brought it home, family members would vomit and throw up blood. They would die if not immediately cured. If the owner of the anting-anting in the tingkep is not home, or comes home immediately, he might not reach his family breathing. The most vulnerable are badly affected. Many of the elders here died.

Pala'wan elder Lumbog Buat also elaborated on how certain mutja carried in tingkep bestow power upon the owner: "According to our ancestors' tale, if they have a mutja that prolongs their lives, they will reach one hundred years before dying. Their other mutja are hidden in the tingkep. In other times, we carry it there while performing [nagba] *basal* [playing of gongs], of course it is worn by us because the mutja are placed in there."

In these ways, the tingkep's character and meaning comprehensively influences Pala'wan life and livelihood across the uplands. Both the exterior of the basket (forest materials, patterns, and colors) and the entities within (powerful forest spirits and supernatural mutja) are subsumed by and entangle the human and nonhuman relations of Pala'wan worlds. Pala'wan describe the tingkep's capacity to bridge human and nonhuman relations through customary rituals and practice, myth and the spiritual realm, and the everyday of upland living. The tingkep's sociomaterial character, meaning, and significance are deeply relational, spanning multiple spheres of Pala'wan life, livelihood, and beliefs, including biopolitical interventions.

Highland Pala'wan have long engaged the productive potential of human and nonhuman relations across diverse forest landscapes to contend with socioecological change and uncertainty over time and space. The sociocultural, ecological, and material dimensions of swidden mosaics converge in the social reproduction, livelihood, and customary practices of Pala'wan families and communities.

This chapter has shown how fluid social relations, mobile swiddens with quick-growing and transplantable crops, and diverse caloric-rich fallows have given Pala'wan degrees of political autonomy. Such spatially complex livelihoods spread risk and reduce vulnerability while also leaving a cultural signature that marks the use of and claims to ancestral territory. As Tsing (2005) notes, the socioecological heterogeneity of swiddens navigates a complex trajectory of relationality between humans' deliberate manipulation of swiddens (e.g., transplanting suckers or keeping root crops grounded) and the emergence of swiddens on their "own way" through assemblages of flora and fauna that influence creatures and customs in the uplands. Pala'wan social relations, livelihoods, custom, swiddens, and forest landscapes work as a rhythmic whole to manage change and generate opportunities (2005).

The tingkep is deeply embedded in and mediates these social relations, rituals, and everyday practices within human and nonhuman worlds. The basket's agency is derived from intersecting human and nonhuman relations and is further formed through various beliefs, ideas, and practices (Miller 2005). Like other customary objects, it is entwined in a web of social relations, objects, and livelihood practices that unite family, kin, forest landscapes, and spirit worlds in the southern highlands.

The tingkep consists of signifiers and characters that have no clear social and ecological boundaries. In the uplands, its sociomaterial character comprises different NTFPs, resin, smoke, ash, and soot. The basket—often soaked in animal blood—may hold tobacco, rice, or powerful mutja talismans. Its woven skin signifies forest natures and superhuman entities; it invokes female forest spirits and mediates complex healing rituals. And yet the tingkep is now also entangled in nonstate reforms. The basket, other customary objects, and associated livelihoods leverage social and ecological reform among the Pala'wan. Nonstate actors have long sought to use the tingkep's deeply personal and intensely public character to access, govern, and unmake Indigenous worlds (West 2006; Li 2007, 2010, 2016).

In the next chapter, I explore how early Spanish colonial law and ideology initiated social and material reform campaigns to upend the human and nonhuman worlds of Indigenous uplanders. I show how Spanish colonial foresters constructed upland peoples in broadly nonstate spaces as heathen, criminal tribal farmers—the ultimate "anticitizens." This partly built a foundation for the American colonial and contemporary biopolitical interventions in the uplands several centuries later.

SPANISH COLONISTS, FORESTS, AND GODS

> To reduce to ashes an extensive and richly populated area sometimes composed of old trees . . . and move the following year to another site with the same operation, is an unspeakable practice whose ravages must be avoided at all costs. This barbaric system takes incessantly from the state, without profit for anyone, an immense amount of timber . . . and hinders the civilization and culture of the Indigenous, favouring their tendency to live isolated and hidden in the woods in miserable huts that serve as a safe shelter for evildoers.

—Jordana y Morera (1874)

Spanish state forester Jordana y Morera's racialized account of swidden farming reveals how colonial biopolitics were already seeking to reform the lives and livelihoods of Indigenous uplanders in the late nineteenth-century Philippines. Various agents of state governance—colonial foresters, land managers, and Catholic priests—pursued ideologies and interventions to racialize, criminalize, and reform Indigenous peoples through a combination of incentives and coercion. Despite more than three hundred years of occupation, the Spanish colonial state's attempts to impose direct rule across the Philippine archipelago (1565–1898) were often partial and fragmented (Paredes 2013). Nonetheless, their sustained ideologies of neglect and disdain drew political attention to uplanders as tribal "anticitizens." In many respects, the enduring failure of the Spanish colonial state to formally recognize customary property rights and citizenship under statutory law further reinforced and induced social difference and marginality in the uplands. This ideological and institutional legacy persisted into the twenty-first century. As Li (2005, xvii) notes, the uplands "have been constituted as a marginal domain through a long and continuing history of political, economic, and social engagement with the lowlands. Marginality must therefore be understood in terms of relationships, rather than simple facts of geography or ecology." Indigenous peoples often took advantage of the friction of distance and terrain by moving deeper into interior spaces to escape the state's purview (Scott 2009), shifting agriculture and residence patterns to evade colonial controls. Ancestral highland regions became places of sovereign refuge (Henry-Scott 1974).

In this chapter, I examine how the Spanish period of colonial forest governance and missionizing practices represented and administered uplanders as tribal "Others" that required disciplining under forest conservation and salvation through religious reform. The colonists were mainly interested in conserving timber for state economic interests and incorporating uplanders as obedient Catholic disciples. This political economy of colonial governance eventually helped reproduce similar biopolitical practices on Palawan Island. Ultimately, the Spanish—and later American—colonizers' fragmented rule in the southern Philippines would create the conditions for Baptist missionaries and, postindependence, environmental NGOs to impose reforms on "forest-destroying" uplanders, who had long evaded state rule and proselytization (see chap. 4).

Colonial-era ethnoreligious hierarchies and agricultural preferences not only informed the construction of Indigenous subjects and the uplands but also determined the dominant forms of governance to ensure that those beliefs, customs, and agricultural practices deemed appropriate were upheld. Spanish colonial authorities treated Indigenous uplanders as "legal minors with less than full rights, inferior status and reduced capacity to own land and participate in political life" (Lund 2023, 1298). A range of essentialized attributes—ethnicity, custom, beliefs, physicality, and place of dwelling—rendered them visible and subject to violent forms of categorization, codification and degrees of administration, but seldom as rights-bearing subjects under the law (Lund 2023). Centuries of colonialism reclassified people and the environment through admixtures of ideology, religion, law, and property, which elevated or reduced peoples' social, material, and legal standing over time (Lund 2023). Rather than ending after independence, the violence of colonial biopolitics simply intensified in the postcolonial period. Uplanders would negotiate an existence at the interstices of legality and illegality (Lund 2021; Dressler 2009).

Spanish Colonialism, Biopolitics, and the Tribal "Other"

The Spanish colonial administration in the Philippines (1565–1898) conflated Indigenous peoples' occupancy and use of upland areas with pagan tribalism in marginal spaces farthest from God and the state. The Spanish rulers constructed exaggerated, racialized caricatures of Indigenous peoples that reified a divide in culture, livelihood, and residence between lowland areas (coastal plains) and mountainous interior landscapes. As Scott (1974, 7) notes, the Spanish regime and the Church "discovered" upland "tribal" peoples based on the ideological "creation of a distinction between lowland and highland Filipinos that contrasted submission, conversion, and civilization on the one hand with independence, paganism and savagery on the other." Race, religion, and agriculture became firmly enshrined in the country's social hierarchies, physical terrain, and broader state–society relations. Christian lowland "native" populations were ultimately classified as productive sedentary agriculturalists more closely aligned with capital, Christ, and the Crown, while those who resisted were categorized as pagan, primitive, backward, "slash and burn" agriculturalists who subsisted—or languished—in the archipelago's hinterland (Dressler 2009).

Spanish colonists attempted to replicate the settlement patterns, agricultural practices, and property rights institutions of the Spanish motherland in the Philippines (Rafael 2000). While it is difficult to determine the extent of Spanish reforms (e.g., commercial agriculture and Catholicism), as ways of life also converged across upland and lowland settings, considerable colonial labor went into producing racialized hierarchies of social difference (Smith and Dressler 2020). The politics behind constructing such ranked difference reflected how colonists envisioned Philippine society would evolve in the eyes of God, the state, and markets (Scott 1974).

Initially, the Spanish rulers replicated the controversial encomienda system from the Americas. It organized, ordered, and disciplined native laborers as vassals (effectively, as Crown servants) under titled landlords, who afforded them protection in return for regular tribute (Anderson 1976; Rafael 2000). The system, maintained until the mid-seventeenth century, was brutally efficient at driving deeper levels of reform and optimizing "unruly" Indigenous populations who clung to pagan faiths and ways of life. As Anderson (1976, 28) writes, "There is no doubt that *encomienda*, particularly the tribute system as it operated in the Philippines, inflicted considerable hardship on the native population. Tribute payments, which was often the root cause for abuses, could take the form of gold, pearls, wax, cotton cloth or mantas . . . and labor."

This feudal subjugation by state officials and clergy aimed to reorganize "the mobile and dispersed nature of native villages [that] tended to obviate Spanish attempts at colonization and conversion" (Rafael 2000, 88). Similarly, the Spanish Catholic *reducciones*, or Christianized settlements, project aimed to overcome more mobile "interior" living and the sociocultural obstacles that it presented to the conversion and sedentarization of pagan uplanders. According to Rafael (2000, 90), the verb *reducir*—and the system itself—denoted the reduction of "a thing to its former state, to convert, to contract, to divide into small parts, to contain . . . to bring back into obedience."

Both the encomienda and reducciones systems were deeply biopolitical. Dispersed coastal and interior plain populations were reorganized and spatially concentrated into enclosures to create newly settled, circumscribed, and managed agricultural communities. The Spanish colonists believed that it would be easier to acculturate and indoctrinate Indigenous peoples if they lived in smaller territories. As political–administrative units, the reducciones demanded that non-Christian peoples be identified, registered, and subjectified to the Crown and to God. Colonial officials and the Catholic Church used these political enclosures and economic systems to reconstitute "the native" as "subjects of divine and royal laws," whereby name, labor, and tribute could be enumerated and disciplined in line with both Catholicism and Spanish institutions (Rafael 2000, 90). Religious conversion and administrative control initiatives were soon implemented for the purpose of reeducating "lost disciples" who continued farming unruly swiddens in the highlands.

These social reforms were brutally comprehensive. Reducciones and similar systems of spatial control incrementally aligned swidden farmers with the religious logics and labor discipline of the clergy and state officials (e.g., tax and tribute). Non-Christian uplanders and lowlanders were required to adopt fixed and ordered agricultural systems using ox and plow to optimize productivity and ensure surplus yields for those in authority. The broader objective ensured that upland peoples and their livelihoods aligned with the rigid boundaries of the emerging cadastral system, individual property rights, and a European sense of labor time—then, as now, equated with "efficiency" and "productivity" (Dressler 2009).

Lowland rural peoples consigned to Catholicism and more sedentary ways were classified as "civilized" Indios. Spanish colonists assigned this derogatory term to those natives with whom they first had sustained political, economic, and religious engagement. They reflected the Spaniards' historical self-image: nascent sedentary agriculturalists who were tentatively faithful to both Crown and God (Hobsbawm and Ranger 2012). Yet as lowland Indio populations were proselytized and incorporated into the colonial administrative fold—amid the gradual deforestation of lowland areas for commercial agriculture, timber production, and colonial infrastructure—other rural peoples broadly resisted colonial incorporation and proselytization by retreating to, or remaining in, the forested highlands of the archipelago. They eventually came to be known as racialized "non-Christian tribes" (Scott 1974).

As the colonial period progressed, the Spaniards became increasingly intent on juxtaposing the mobile and destructive character of upland swidden practices with the productive, sedentary properties of irrigated rice cultivation in the lowlands (Dressler 2009). In time, such social, ecological, and geographical differences emerged as a naturalized and enduring binary in Philippine society and a central facet of the state's policy architecture for governing forests, lands, and peoples. Swidden farmers were regarded as mobile pagans who razed valuable state forests. Contemptuous, racialized terms denoted Indigenous high-landers' lowly evolutionary status (e.g., *tribus infieles, salvajes,* and *indenas*) and subjected them to enduring civilizing discourses that were predicated on the rule of a Christian God and the colonial state.

The Spanish colonists, informed by European ethnologists' preoccupation with classifying race and population (Mojares 2013), normalized a biologically deterministic racial hierarchy of Indigenous subjects in terms of "three waves of migration" (Aguilar 2005, 606). The theory posited that each subsequent wave of native migration to and across the archipelago was more advanced and civilized than the last and pushed earlier arrivals deeper into the mountainous uplands (Aguilar 2005, 606). The first wave comprised a primordial stock of short, darker-skinned "Negrito" peoples—a "distinct race"—who occupied or were eventually displaced farther into the hinterland by a second wave of invading Malays. This second wave of Malays, who initially settled in coastal areas, were subsequently driven into the interior by a third wave of more advanced Malays, the Indios, who eventually occupied and dominated the coastal lowlands (Aguilar 2005). The Spanish clergy and administrators "mapped and marked" the first and sec-ond waves of native arrivals based on their residence (mountainous, interior forested areas) and their assumed sociocultural characteristics (tribal, animist, recalcitrant, and resistant to Christianization; Aguilar 2005). For instance, Padre Chirino, a Spanish Catholic missionary who traveled to Panay Island in 1592, asserted:

> Among the *Bisaya* [Indios], there are also some Negroes. They are less black and ugly than those of Guinea, and they are much smaller and weaker, but their hair and beard are just the same. They are more barba-rous and wild than the *Bisayas* and other Filipinos, for they have neither houses nor any fixed sites for dwelling. They neither plant nor reap, but live like wild beasts, wandering with their wives and children through the mountains, almost naked. They hunt deer and wild boar, and when they kill one, they stop right there until all the flesh is consumed. Of property they have nothing except the bow and arrow. (Chirno 1592, cited in Barrows 1924, 4)

Such racialized constructions further influenced the management directions of the colony's first formal forestry department, Inspección General de Mon-tes, in 1863 (Bankoff 2004; Dressler 2009; Smith and Dressler 2020). As Spanish colonists increasingly viewed upland forests as sources of commercial timber for shipbuilding, city reconstruction, and timber export (Bankoff 2004), they fur-ther criminalized the darker-skinned swidden farmers through ideological, legal, and scientific rationale. Most upland farmers were perceived as unproductively and wastefully using lucrative state timber and forest resources (Dressler 2009). For the colonist, the outwardly destructive and primitive character of swidden agriculture (*caiñgin*, from the Tagalog, *kaingin*) was anathema to colonial and religious ideals of productivity, pacification, and indoctrination (Bankoff 2013; Olofson 1980).

Ultimately, tribalized swidden farmers were cast as anticitizens. The extensive rotational character of slashing and burning mature forests for "temporary" farms became broadly untenable for Spanish colonial foresters between the 1870s and 1890s, despite laws protecting Indigenous peoples' rights to harvest timber for their own use. Such contempt toward swidden was strikingly evident among Spanish foresters within the Inspección General de Montes. For example, Ramon Jordana y Morera (1874, 29) wrote with anxious concern about swidden practices to others in the forestry unit: "[This] . . . barbaric farming systems takes incessantly from the state, without profit for anyone, an immense amount of timber . . . and hinders the civilization and culture of the Indigenous peoples, favoring their tendency to live isolated and hidden in the forests in miserable huts that serve as a safe shelter for evildoers."

Jordana y Morera (1879, 23) lamented further that the state's limited reach to manage tribal uplanders was allowing slash and burns to expand, to the detriment of state forests: "Due to the impossibility of concentrating administrative action in those localities due to a lack of personnel, the effect of burns or *cainges* [kaingin/swidden] from Indigenous people who run without the slightest anticipation, leads to overflowing rivers that produce in some points, during the rainy season, great floods that destroy houses and crops."

Another colonial botanist, Vidal y Soler (1874, 40), described how Moorish (Muslim) kaingin was destroying valuable state timber on the island of Mindanao, further illustrating how Spanish colonial discourses conflated and reinforced forest degradation, swidden, and sociobiological difference:

> In all the islands of the South, considerable quantities of good timber are lost that would be used for construction. The Moorish [Muslim] farms move frequently, and they burn the forest in the place they have chosen to grow their crops for a couple of years. And the aboriginal Indians, always fleeing from their contact, are like their forerunners taking fire into the interior [of the forest]. This matter demands a preferential interest on the part of the local authorities, whose zeal should excite their superiors.

These published chronicles of Spanish foresters and botanists politically conflated the primitive character of upland farmers and destructive swidden with rampant deforestation and degradation, which threatened state forest assets.

The political production of such cultural difference formed the basis for explicit social and geographical differentiation over several centuries (Smith and Dressler 2020).

In 1874, Inspección General de Montes declared swidden illegal in the public domain (i.e., on Crown land; Dressler 2009) and, in 1889, implemented a set of "Definitive Forest Laws and Regulations" imposing steep fines for any violations (Scott 1979, 59). Much of this enforcement was based on the Regalian Doctrine: a principle that claimed all lands (*de dominio publico*) not registered as private property as vested in the Crown (Noblejas and Noblejas 1992), legally denying uplanders formal tenure on ancestral lands. Crown and state ownership were now broadly synonymous. The Spanish colonial government had full control over forests, water, and mineral resources in the public domain and limited Indigenous peoples' rights to ancestral lands and resources (Lynch 1982, 274). The state's denial of formal tenure rights for Indigenous uplanders persisted for centuries despite the Spaniards' Law of the Indies, which afforded partial recognition to Indigenous conceptions of (usufruct) ownership through occupancy, clearing, and cultivation of forestlands.[1] Upland peoples who resisted incorporation into the tribute system and fixed-plot sedentary agriculture were denied the opportunity to claim land under formal title (Dressler 2009).

Religion, Trade, and Territory in Southern Palawan

Together, the Spanish colonial state and Catholic Church soon maintained broad political control and management of the more accessible lowland peoples, lands, and agriculture. Yet many uplanders avoided proselytization and state pressures by seeking refuge on ancestral lands within mountainous interiors (Scott 1974). For centuries, the highlanders of southern Palawan largely evaded colonial rule and proselytization.

In the early days of colonial rule, the Spanish colonial presence was most pronounced in the northern and central parts of Palawan (Eder 1999, 22) and manifested through missionizing campaigns in 1622–23 (Ocampo 1985). In 1622, the Bishop of Cebu, Friar Pedro Acre, and the governor-general of the Philippines authorized the first missionary work on Cuyo Island (off Palawan's mainland), allowing Augustinian Recollects to proselytize some of the island's Indigenous inhabitants through tribute, resettlement, baptism, and indoctrination (Ocampo 1985, 23). Ocampo (1985, 24) writes that the missionaries "were well received by the natives, except the native priests [babalyan or beljan]," who feared losing their customary authority if the friars were permitted to "carry out the *reduccion* [gathering] of households in a central place, building a church in the middle of it, and spread the doctrine of Catholicism" (24). He notes further that, "after several months of persuasion, they were able to baptize a thousand natives" (24). The next year, the Spanish church and colonial administrators replicated these efforts on mainland Palawan with mixed success. Despite deploying soldiers, sending additional friars, and establishing a stone fort at Taytay, initial attempts to "encourage the natives to gather" were "strongly refused"; according to the friars, the "natives" "could not leave their old way of life and feared [that by being concentrated in settlements] they would be very open to attack, especially by the Moors" (24). These events signified early, episodic Indigenous resistance to colonial biopolitical interventions.

The Spaniards exerted a combination of military force and religious power in their attempts to colonize the remainder of Palawan from the northern administrative centers (Ocampo 1985). Proselytization efforts unfolded recursively on the island owing to the lack of financing and the deployment of religious and military staff to Zamboanga in response to the threat of Chinese and Muslim control. In other cases, local mayors exploited Indigenous peoples (likely Tagbanua) to intensify the harvest of lucrative forest products, such as beeswax (and presumably honey), "forc[ing] [many] to evacuate to remote areas or the mountains" and making "evangelization difficult" (27) in the lowlands. Missionizing was eventually reinvigorated as church officials advocated for deeper proselytization of Indigenous leaders and their children from the northern to the central areas of the island; friars "patiently taught [Indigenous children] reading, writing, arithmetic and music" so that they gained the "qualities necessary for them to govern their respective peoples" as part of reform initiatives (27). In southern Palawan, Spanish political control and religious influence had even less traction among recalcitrant Muslim peoples along the coast and Pala'wan in the highlands (Ocampo 1996; Warren 2007).

The powerful sultanates of Maguindanao, Cotabato, and Sulu—reflections of the region's Islamization from the late 1300s—retained control of southern Palawan through the influence of religion, trade, and tribute, and the threat of raiding and enslavement (McKenna 1996). Notably, the Sulu Sultanate's political influence and territory encompassed the Sulu archipelago, northeastern Borneo, and southern Palawan from 1778 to at least 1894 (Warren 2007; Ocampo 1996, 30). At that time, the Sultanate-affiliated Jama Mapun and Tausug people established well-populated trading settlements in the coastal lowlands of southern Palawan, further restricting the Pala'wan to the uplands (Warren 2007; Smith 2015). Unlike the proselytization efforts exerted by friars further north, trade, tribute, and coercion between the lowland Muslim and highland Pala'wan (and

Tagbanua) spread the influence of Islam, its hereditary leadership structure, and material culture into the mountainous interior (Ocampo 1985).

The lowland Tausug and Jama Mapun and Pala'wan highlanders emerged as key conduits within the so-called Sulu Zone trade network, which connected the Sultanate, the British, and Chinese merchants (Warren 2007). In particular, the Tausug drew heavily on Pala'wan labor to acquire marine (e.g., *tripang*) and forest products (e.g., edible birds' nests and rattan) to trade with the British, who then sold the products to Chinese merchants for sale in Asia. Crucially, they also sourced significant amounts of tribute rice from upland swiddens (and possibly lowland paddies) to be "shipped to [rice poor] Jolo in large quantities" (Warren 2007, 99).[2] In exchange for tribute rice and nontimber forest products (NTFPs), Tausug provided the Pala'wan (and other Indigenous uplanders across the island) with material goods of increasing cultural importance for upland living and ritual, including salt, "cloth, porcelain, ironware, bronze gongs, and other items of Bornean and Chinese trades" (Ocampo 1996, 27; Dressler 2009). The spread of Malay/Javanese (Islamic) lowland material culture into the forest interior was particularly pronounced, with Pala'wan and Tagbanua wealth being measured by the number of gongs, kris swords, and jars that a family held in their home (Ocampo 1985; Fox 1954). Ritual paraphernalia (e.g., bean boxes), ceremonial art, basic motifs, and inscriptions on kris swords, textiles, and carvings were similar in character and origin.

Indigenous uplanders lived in constant fear that their children and women would be captured and enslaved by the Tausug in retribution for defaulting on tribute payments to the Datus who controlled the upland–lowland trading systems (Ocampo 1985; Fox 1954). Persistent Tausug attacks and harassment were carried out against those uplanders and colonists who were perceived as a threat to their trade, wealth, and religious influence across the south of the archipelago. Ultimately, however, the Spanish failed to significantly reform Pala'wan uplanders' pre-Islamic (proto-Malaya) customary beliefs and practices, or the mainstay of swidden agriculture, in the south. The influence of Islam from Tausug and other Muslim groups was more pronounced within Pala'wan social and material relations. Unlike the island's state-governed and Catholicized north, Pala'wan in the southern uplands retained greater degrees of social autonomy and sovereign territories.

Spanish naval forces only temporarily subdued the Muslim influence in the south. Indigenous and Muslim lowlanders repeatedly breached colonial infrastructure, and most military forts and agricultural stations in Balabac and Puerto Princesa were eventually abandoned (Ocampo 1996, 29, 32–33).[3] The Filipino revolutionary resistance in 1896, the loss of the Spanish-American War in 1898, and the arrival of American colonists in Manila considerably weakened Spain's claim to the country. The US military presence expanded across the country, suppressing Filipino and Muslim resistance in the brief and bloody conflict known as the Philippine-American War (1899–1902). After the US military suppressed the insurrection, the Americans built upon Spain's colonial architecture to further develop and inculcate an expansive biopolitical governance regime.

Despite the lengthy period of Spanish rule, the upland peoples of southern Palawan—and the inhabitants of other less governable places—proved to be beyond the reach of the Catholic Church and Spanish colonial state. In highland areas, the Pala'wan and other self-governing groups were ultimately left to the long-standing trade relations with lowland Muslims and the sultanates who had dominated the region for considerably longer than the Spanish. Nonetheless, the racialization of Indigenous peoples as uncivilized, forest-destroying tribal uplanders that had become entrenched in Spanish colonial society and forest policy would endure for centuries.

In outer island areas, Spanish colonizers and missionaries had limited success in reforming and optimizing Indigenous uplanders into modern, governable Catholic subjects. It was largely Indigenous lowlanders, near Spanish outposts and religious institutions, who became (incompletely) incorporated into the administration's reform agenda. The Muslim sultanates and their intermediaries were more successful in influencing Indigenous uplanders through a system of raiding, trade, and tribute. As chapter 4 will show, the American colonial regime's attempts at biopolitical reform and incorporation of tribal uplanders were better organized and far reaching. Despite this, neither the American administration (military or civilian) nor the Church had significant or lasting effects on uplander life and livelihoods in southern Palawan. Consequently, the door to new nonstate biopolitical reforms was left open.

I explore next how efforts at biopolitical governance intensified under the American occupation and after the independence of the Philippines in 1946. Baptist religious fervor and—many decades later—environmental NGOs would reach deeper into the mountains of southern Palawan to monitor, regulate, and optimize uplanders' residence patterns, livelihoods, and social behaviors. In time, a mix of Spanish and American colonial laws, policies, and practices would underpin contemporary state and nonstate reforms. The Pala'wan highlands served as the biopolitical testing ground.

4

AMERICAN FORESTERS, NONSTATE RULE, AND THE TRIBAL "OTHER"

> It is evident that, if we are not to fail in our duty toward the savage or half-civilized Philippine peoples, active measures must be taken for the gathering of reliable information concerning them as a basis for legislation, and an act has therefore been passed by the commission creating a bureau of non-Christian tribes.
>
> —Philippine Commission (1901)

The American (United States, US) colonial regime in the Philippines reinforced the racialized social differences inherited from Spanish rule and helped set the biopolitical architecture for governing the uplands in the postcolonial period (Ileto 1979; Dressler 2009).[1] From 1898 to 1946, an interim US military and subsequent civilian governments reproduced the long-standing colonial practices of the Spanish by further codifying and ordering segments of the rural population to align with social hierarchies, land use preferences, and religious ideals. After Philippine independence in 1946, an influx of Baptist evangelicals from the US and other NGOs harnessed and often bypassed the state's bureaucracy and governance reach into remote frontier regions. Nonstate actors soon filled governmental institutional voids, aligning Indigenous beliefs and livelihood behaviors with markets, private property, and sedentarism through newer practices of biomedical monitoring, ecological controls, and social incentives. In time, environmental NGOs offered the benefits of de facto citizenship—social contracts, recognition, and Indigenous rights—to uplanders who were willing to present themselves as sufficiently traditional and sustainable. In contrast, evangelical beliefs, strictures, and incentives sought to dismantle aspects of indigeneity so often valorized by other civil society organizations.

This chapter proceeds in two parts. The first part analyzes how the early period of the American colonial regime's administration of modern hygiene, beliefs, and livelihood aligned with individual control, responsibility, and ownership of body and property. Proselytizing and missionary activities took a Protestant turn during the American colonial period, which was more in keeping with

88

these reformist ideals. The US administration created new bureaucracies to govern Indigenous subjects accordingly and instituted the country's first comprehensive census—a key biopolitical instrument of the new colony—to formalize and entrench the racialized subject categories of Spanish rule. Like their Spanish predecessors, American colonial forestry officials continued to campaign against swidden agriculture as detrimental to both valuable state forestry assets and efforts to civilize the "non-Christian tribes." The American regime rendered the racialized social hierarchy explicit through imperial hubris and administrative codification, entrenching the minority status of uplanders for decades to come.

The second part of the chapter considers how the American regime's biopolitics supported and opened the door to (Baptist and Adventist) evangelical religious ideals and influence in the uplands. With few political obstacles in their way, evangelical missions expanded and intensified the colonial regime's civilizing doctrine and logic into remote corners of the country. Yet, despite (or perhaps because of) these efforts, Indigenous peoples continued to use material objects with considerable syncretic powers—often drawn from Catholic mysticism and supernatural deities—to resist colonial practices of erasure.

The section then examines how, postindependence, newer global forms of state and nonstate recognition of indigeneity discursively bureaucratized, reified, and reformed uplander identity, customs, and livelihoods around tradition and sustainability. I explore the governance tensions and contradictions that emerge with attempts to construct and deconstruct modern imaginaries of customary peoples, old growth forests, and ungovernable uplands. As Tsing (2005, 158) notes, "the residents of the uplands . . . have come into a new visibility." In this (nonstate) governance mix, they may be represented as either "closed and static repositories of custom and tradition" or in need of becoming "individualistic entrepreneurs with no commitments to local social life or culture" (159). This holds for the uplands of southern Palawan, where "a field of attraction must be created to nurture and maintain the relationship between the rural community and its experts" (160). This discursive field of attraction emerged through the early social histories, practices and impacts of nonstate actors in the Philippines.

The American Colonial Period

The Tribal Anticitizen

American colonists in the Philippines set out to render the fluid trade networks, residential patterns, and agricultural mosaics of upland populations into discrete administrative categories and divisions. Within these divisions, they sought to sedentarize and civilize uplanders through a violent admixture of Christianity, hygiene, and education (Anderson 2006). The Bureau of Non-Christian Tribes, established in October 1901 within the Department of the Interior, was the first to classify and administer Indigenous uplanders as anticitizens without religion. Its primary objective was to obtain ethnological information about "wild tribals" to better understand, monitor, and manage them (Philippine Commission 1903). Dean Worcester, then head of the Department of the Interior and an amateur ethnologist, led the charge to catalog so-called non-Christian tribes across the hinterlands of the newly acquired colony. Under chief David Barrows, the bureau hoped to overcome a "lamentable lack of accurate information as to the non-Christian tribes of the Philippine Islands" attributed to the previously deficient Spanish administration (Philippine Commission 1901, 35). The limited staff set out to conduct predesigned, "systematic investigations . . . of non-Christian tribes of the Philippines Islands" (Act No. 253, 1901, cited in Hutterer 1978, 142). The bureau's findings—categorized into number of tribes, territories, languages, beliefs, and hygienic practices—ultimately flattened and simplified the complexity and diversity of Indigenous uplanders to make administrative assimilation more efficient and effective (Blanco 2009). Such census-making underpinned the early biopolitics of state-building processes.

The bureau's far-reaching civilizing mission involved "the advancement of the non-Christian elements of our populations to equality and unification with the highly civilized Christian inhabitants" (Kalaw 1919, 2). Under the Jones Law, the bureau strove to assimilate the "tribal Other" by incorporating their residential patterns, livelihoods, and social practices into the lowland agrarian political economy. This included a "closer settlement policy whereby people of semi-nomadic race are induced to leave their wild habitat and settle in organized communities . . . [attend] public schools . . . and the system of public health" (Kalaw 1919, 3). A growing labor force of Filipino civil servants—including teachers, doctors, and veterinarians—soon worked in the US colonial administration to further facilitate the assimilation of Indigenous uplanders living in remote regions of the country.

The bureau used its newly cataloged demographic data to establish the 1903 census, which officially categorized rural Philippine subjects into "civilized and wild" (see figs. 4.1 and 4.2). The census was a rather blunt instrument, categorizing Christian lowlanders as peoples with "considerable degrees of civilization" and non-Christian "tribal" uplanders as "wild and uncivilized," arguing that the "Philippine census is somewhat simpler, the difference being due to the more homogenous character of the population of the Philippine Islands" (Philippine Commission 1903, 9). As figure 4.2 shows, the census classified 91.5 percent of the population as civilized and 8.5 percent as "wild," with the latter group consisting of racially homogenized "tribal" labels such as "Moros" or "Negritos" (Philippine Commission 1903, 15). In crude racialized language, the census definitively

With the exception of the Negritos and the people of foreign birth, all the inhabitants of these islands are believed to be Malays.

These people were, at the time of the census, distributed in 39 civil provinces; the subprovince of Marinduque, which is here given separately, though a part of Tayabas; 9 military districts or comandancias, and 1 city—Manila. They were scattered over 342 islands.

The following table shows, by provinces and comandancias, the total population and its division between the civilized and wild people:

PROVINCE OR COMANDANCIA.	Total population.	Civilized.	Wild.
Cebú	653,727	653,727	
Iloilo	410,315	403,932	6,383
Pangasinán	397,902	394,516	3,386
Leyte	388,922	388,922	
Negros Occidental	308,272	303,660	4,612
Bohol	269,223	269,223	
Sámar	266,237	265,549	688
Batangas	257,715	257,715	
Albay	240,326	239,434	892
Ambos Camarines	239,405	233,472	5,933
Cápiz	230,721	225,092	5,629
Pampanga	223,754	222,656	1,098
Bulacán	223,742	223,327	415
Manila city	219,928	219,928	
Negros Oriental	201,494	184,889	16,605
Ilocos Sur	187,411	173,800	13,611
Ilocos Norte	178,995	176,785	2,210
Misamis	175,683	135,473	40,210
Cagayán	156,239	142,825	13,414
Tayabas[1]	153,065	150,262	2,803
Rizal	150,923	148,502	2,421
La Laguna	148,606	148,606	
La Unión	137,839	127,789	10,050
Tárlac	135,107	133,513	1,594
Cavite	134,779	134,779	
Antique	134,166	131,245	2,921
Nueva Ecija	134,147	132,999	1,148
Cottabato[2]	125,875	2,313	123,562
Sorsogón	120,495	120,454	41
Surigao	115,112	99,298	15,814
Zambales	104,549	101,381	3,168
Isabela	76,431	68,793	7,638
Lepanto Bontoc	72,750	2,467	70,283
Dávao[2]	65,496	20,224	45,272
Nueva Vizcaya	62,541	16,026	46,515
Romblón	52,848	52,848	
Abra	51,860	37,823	14,037
Marinduque[3]	51,674	51,674	
Joló[2]	51,389	1,270	50,119
Bataán	46,787	45,166	1,621
Zamboanga[2]	44,322	20,692	23,630
Masbate	43,675	43,675	
Mindoro	39,582	32,318	7,264
Basilan[2]	30,179	1,331	28,848
Paragua	29,351	27,493	1,858
Siassi[2]	24,562	297	24,265
Dapitan[2]	23,577	17,154	6,423
Benguet	22,745	917	21,828
Tawi Tawi[2]	14,638	93	14,545
Paragua Sur[2]	6,345	1,359	4,986

[1] Exclusive of subprovince of Marinduque.
[2] Comandancia.
[3] Subprovince of Tayabas.

FIGURE 4.1. Racialized Philippines Commission census statistics, 1903. Source: Philippine Commission 1903, 16.

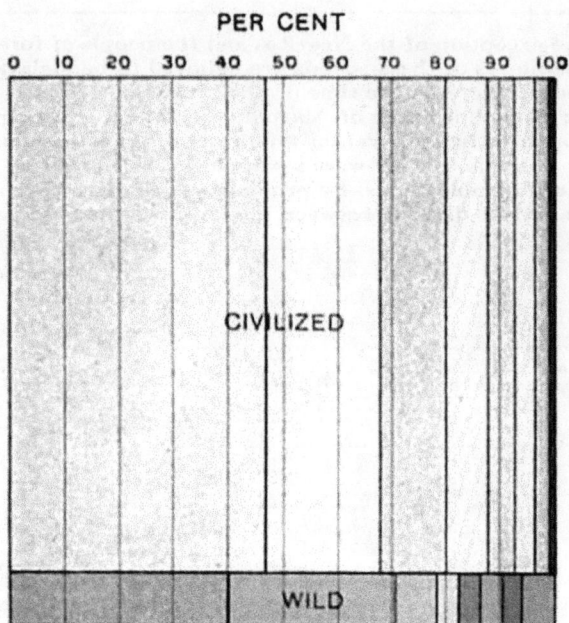

TOTAL POPULATION

DISTRIBUTED AMONG THE CIVILIZED AND THE WILD TRIBES

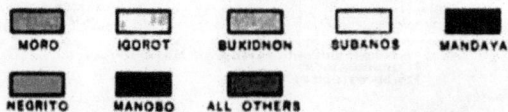

FIGURE 4.2. The "civilized" and "wild" tribes of the inaugural Philippines census, 1903. Source: Philippine Commission 1903, 16.

registered uplanders as tribalized "Others." Colonial politics of difference were now officially codified in the state's—and soon society's—biopolitical architecture.

The coding categories of the census forms reflected the reified social, economic, and biophysical attributes (i.e., citizenship, education, literacy, school attendance, family size) most commonly associated with sedentary, fixed-plot, "civilized" Christian farmers residing in the lowlands. In many ways, the racialized representation of so-called pagan uplanders already ensured that they would fail to align with these formalizing attributes (see fig. 4.2; Philippine Commission 1903), as Barrows espoused with dehumanizing language, writing: these "little savages excited the attention of the first Spaniards [asking] "whence came these curious little people, and what does their presence here signify? . . . The Aetas of the Philippines are not the only black little dwarfs in the world" (Barrows, 1926, 5).

The census's racialized social categories influenced the governance of Indigenous uplanders over several subsequent decades, particularly the Bureau of Forestry's targeted campaigns to reform uplanders' livelihoods and the apparent threat they posed to forest landscapes.

Criminalizing Swidden, Optimizing the Tribal Uplands

The American colonial administrations' civilizing missions built on and invested in varied legal practices to comprehensively manage and optimize the forests and peoples of the Philippine uplands. The colonial regime upheld the essence of Spanish property rights law, reinforced individual property rights, and continued to partition the forest commons according to race, religion, and elevation. The Americans cited res nullius to uphold the Regalian Doctrine's myth of unoccupied and uncultivated lands, erroneously reaffirming highland spaces as state lands or timberlands. By default, all uplanders became "illegal" occupants and users of their own ancestral lands (Dressler 2009). Later, the Philippine Bill of 1902 converted Spanish Crown land to public (state) lands using the Torrens system of land registration and transfer, which titled plots to individual owners and rendered land an exclusive, ownable, and bounded commodity (Dressler 2009). The bill ensured that Indigenous lands could be claimed, bought, and sold by others as private title in the lowlands. Unproductive "slashing and burning" was condemned.

In this discursive space, rural lowland farmers with claims to private title were represented as productive and legitimate citizens, subject to taxes, regulations, and degrees of recognition from the colonial government. Conversely, upland farmers who lacked formal, state-sanctioned land claims were often denied recognition and rendered invisible by the state. As Lund (2021, 8) notes, property amounted to "a legitimized claim to something of value sanctioned by some form of public authority . . . Struggles over property [could] therefore be seen as struggles for the recognition of a wide variety of rights to access resources in various ways." The US colonial administration's need for intact timber forests and entrenched prejudices toward Indigenous swidden lifeways legitimized efforts to promote fixed-area tenure to induce sedentarism. Colonial forestry officials aimed to resettle uplanders on temporary, conditional tenurial instruments and evict those Indigenous occupants of state lands who lacked formal tenure.

The Bureau of Forestry, established in 1900, drew on scientific forestry methods to manage public lands and timberlands under colonial state jurisdiction and aimed to maintain forest cover for timber and territorial control by regulating and sedentarizing swidden farmers. Inaugural bureau chief, Captain George Ahern, aimed to professionalize the relatively well-funded agency's biopolitical strategies (Nano 1939, 22). Ahern—a Yale-educated US military officer—hired American (and eventually Filipino) foresters to undertake forest mapping, inventories, and the delineating of forest reserves to achieve the alleged maximum sustained yield of timber harvests (Pinchot 1903). The bureau considered scientific stand management and sawmilling to be the most efficient approaches to optimizing timber extraction and revenue for American colonial prosperity (Dressler 2009). The prioritization of high-value species (e.g., Narra), timber reserves, and sustained yield principles contrasted starkly with the perceived irrationality and destruction by swidden uplanders.

By 1916, the US administration began to facilitate the expulsion of "illegal occupants" from the uplands to protect timber stands (Roth 1983; Bagadion 2000). Bureau of Forestry staff increasingly expressed deep disdain for swidden farmers, particularly repeat burning for field preparation. Henry Whitford (1921, xx)—a forester working for the Bureau of Forestry's Division of Investigation— repeatedly demonized farmers' use of fire in countless bureau reports and other official correspondence:

> As a matter of fact, fires are uncontrolled, the control is centered on maintaining virgin forest areas, but in spite of efforts of forest officers, large areas of forests are still being destroyed. The fact of the matter is that this primitive method of agriculture has dominated nearly three-fourth of the virgin forest areas of the Philippines. Strange to say, with their crude instruments they are better able to conquer the forest than rid the soil of the rank growth of grass with its network of underground stems which fire will not kill. Here is a case where grass and not forests has dominated primitive mankind.

Bureau foresters believed that fire had the potential to quickly decimate timber stands and impede the regenerative capacity of forest landscapes. The successive clearing and burning of swidden plots were known to destroy saplings and the suffocating growth of expansive *Imperata cylindrica* (cogon) grasslands suppressed forest regrowth across landscapes (Smith and Dressler 2020). While foresters saw this fast-growing, flammable grassland as an idle, degraded wasteland, resourceful uplanders used the grasses extensively for weaving, thatching, insulation, and fodder (Dove 1986; Montefrio and Dressler 2016). Then, as now, state foresters undertook intermittent policing and monitoring efforts to curb cogon fires—the principal threat to the fragile saplings used in reforestation and regeneration (Smith and Dressler 2020).[2]

The bureau's broader conviction that expansive swidden practices obstructed the administration's civilizing mission (and extractive interests in upland forests) led to increasingly rigid and comprehensive antiswidden laws. In an overview of colonial forestry laws governing swidden (kaingin), Jose Nano describes how the Kaingin Law of 1900 prohibited swidden and authorized bureau staff to issue fines of up to US$100 or a thirty-day prison sentence or both (Nano 1939, 30). A first offense by "non-Christian tribes" was "dismissed with a warning" (Bureau of Forestry 1932, 125), but the second offense led to a tripling of fines and several months' imprisonment (according to the prewar government's Commonwealth Act No. 447; see also Section 25, Bureau of Forestry, 1904, 15).[3]

In more accessible areas, the bureau proactively collected information and evidence to convict swiddeners in upland "public forests." Foresters now invested in extensive patrols, taking detailed notes, and sketching maps of the forests destroyed by swidden. They also interviewed other "non-Christian subjects" who witnessed the clearing and burning of forests (Bureau of Forestry 1932, 197), requesting the offender's name, place of residence and the location of the "swidden violation" for potential future conviction (116, 280–85). In most cases, foresters targeted farmers who cleared valuable hardwoods, such as tindalo, akle, and molave (*Vitex parviflora*), without a permit, noting the height and perimeter of the cleared trees and the fine levied (Bureau of Forestry 1904, 15; Bureau of Forestry 1932, 199). However, even this meticulously planned antiswidden panopticon had deficiencies.

The bureau (rather haphazardly) aimed to register, license, and control swidden farming through a permitting system adopted from the 1904 Forestry Manual. Informed by the Spaniards' licensing system, this plan drew on American settler ideals of productivity, private property, and capital to satiate the uplanders' supposedly suppressed desires for sedentary livelihoods (Bureau of Forestry 1901). Colonial foresters issued swidden permits for areas that were deemed better suited for agriculture than timber production to keep the timber concessions free of unruly kaingin (Section 25, 45).[4] A kaingin licensing system was thus established to administer and centralize swidden farmers across the country. Successful applicants were granted individual ownership over a plot of land. However, this classification and codification system was predisposed to failure, since most swiddeners cleared, fallowed, and rotated their fields annually across remote and rugged terrain (Bureau of Forestry 1900, 12; Bureau of Forestry 1904). Once sufficiently fallowed into older forest, family claims to swidden lands attenuated as they reverted to the forest commons.

American foresters continued to use the need to govern recalcitrant uplanders and the protection of valuable timber as pretexts for the sustained racialization and criminalization of uplanders whose access to and use of state forests should be restricted. More than half a century later, the very same colonial ideologies and practices would spatially manifest as overlapping zoning systems in which nonstate actors launched their own projects of reform, supporting or bypassing state law and policy in highland areas.

American Colonial "Civilizing" Policies in Southern Palawan

Internal migration to Palawan increased considerably for thirty years as the American colonial administration consolidated peripheral islands into what became the Palawan Province (Ocampo 1996). From 1910 onward, farmers from Cuyo, the Visayas, and Luzon sought to escape civil unrest and economic decline by relocating to and availing themselves of Palawan's fertile lands, bountiful fishing grounds, and relative safety (Eder and Fernandez 1996). As families stayed, cultivated lands, and invited others, the population of Palawan proper nearly doubled in fifteen years, increasing from 35,369 to 69,053 between 1903 and 1918 (Eder 1999, 22). While the island's central and south-central regions experienced considerable in-migration and settlement from these regions, the south remained only sparsely populated with Muslim settlers and, gradually, Christian lowlanders (Macdonald 2007). The Pala'wan mingled with both groups in lowland areas but largely resided within the southern ranges.

In central Palawan, the American colonial administration attempted to incorporate the Tagbanua (and likely some Pala'wan) into lowland lifeways, particularly fixed plots, commercial agriculture, and Christian ideals. The Bureau of Non-Christian Tribes partly repurposed the Spanish reducción program, developing rancherias to accommodate the induced settlement of Indios (Bayuga 1989). For instance, in 1910, the provincial colonial governor of Palawan, Edward Miller, constructed an agricultural school in Aborlan—the heartland of Tagbanua society—to tame a "savage stock of tribal natives" (Bayuga 1989, 6). The Tagbanua were confined to the school's "reservation," where they learned to build permanent houses, follow a strict regimen, and understand the value of time. They were also forbidden from speaking their dialect (1989). Anyone attempting to escape to the uplands was punished. Such biopolitical efforts to discipline were extended further south to Bataraza, where Governor Miller aimed to pacify "restive" Moros and "members of non-Christian tribes [by having them] take up their residences at places designated by him, if such a course [was] deemed to be in the interest of public order" (Department of the Interior 1910, 12). The "worst of these settlements" would be brought to and governed under the territorial jurisdiction of Datto Bata-rasa (Department of the Interior 1910, 12).

Even as northern Palawan was settled, the American colonial administration's hold in the southern lowlands remained fragmented and largely absent from the Mantalingahan mountain range. Muslim Tausug, under the Sulu Sultanate's authority, retained political control over the coastal and inland lowlands, sustaining social influence by collecting tributes of rice, forest products, and labor for the Sultan's datu into the early 1900s (Warren 2007, 71). In contrast, Pala'wan living deeper in the interior were subjected to tribute collection on a much smaller scale and likely less influenced by the Tausug (Brown 1991, 27).

American naval power gradually suppressed the Sulu sultanate's regional trading regime and political authority, and Tausug Muslim elites' hold over lowland areas became increasingly tenuous around World War I (Smith 2015). However, the in-migration of Christian settlers remained slow and intermittent, as the Tausug continued to frustrate European and Chinese traders attempting to form upland–lowland trade networks (Smith 2015). As a result, Muslim settlers and Pala'wan remained the two main population cohorts in the Brookes Point lowlands and much of southern Palawan at the time (Brown 1991).

Postindependence In-migration, Land Use, and Pala'wan Marginalization

In the 1950s, after independence, more Christian migrants from the Visayas and Luzon began to settle in Brooke's Point and the newly formed southern municipalities of Rizal (1951) and Bataraza (1964). Over time, they began to "set up a political system that resembled the one they were familiar with in their areas of origin and . . . replace[d] the prevailing [Muslim Tausug] system in southern Palawan" (Brown 1991, 5). The migrants comingled and traded with Muslims in the lowlands and with Pala'wan in the uplands (Dressler 2009). These pioneer migrants initially relied on the Pala'wan for subsistence, but gradually cleared lowland forests for incipient swiddens to cultivate rice and root crops on flat, fertile alluvial lands (Brown 1991; Smith 2015). Trading relied on each group's relative comparative advantage: migrants exchanged lowland goods, such as medicine, fish, and salt for Pala'wan staples, such as cassava, swidden rice, and forest honey (Brown 1991; see also Peterson 1978; Dressler 2009).[5] Such highland-lowland trading practices persist to this day, though increasingly mediated by NGOs and certain state agencies.

In the 1960s, most forests cleared for swidden along the island's southeastern coast were converted into extensive irrigated paddy rice systems (*basakan* in

Cebuano) and permanent homesteads (Macdonald 2007; Dressler 2009; Smith 2015). Migrants drew on skills from their places of origin to expand (initially rainfed, *tubigan*) basakan by clearing forests and amassing flat or undulating lands, which they connected to rudimentary irrigation systems fed by canaled and dammed streams from upland forests. Migrant paddy farmers soon claimed land from the Pala'wan and, exploiting Barangay political networks, demarcated and registered their paddy fields as titled property (Brown 1991).

Migrants relied on their own social relations and exchange networks to manufacture or purchase the capital and infrastructure required to expand basakan production. They gradually required more Pala'wan labor to sustain surplus production. Pala'wan were taken on as iterant day laborers and paid with low daily wages or credit advances in the form of rice and other goods. This burdened Pala'wan laborers with debt and debt-bondage to the migrant landowners engaged in copra and paddy rice production (Brown 1991; Smith 2015).[6] The Christian settlers' claims to vast agricultural lands now provided the first legal means of dispossessing the Pala'wan from lowland areas that had long been used by their ancestors. These lowland incursions were the most recent in a historical series of dispossessions of land and resources that had pushed the Pala'wan further into the interior.

By the 1970s, the populations of Brooke's Point and Bataraza proper had gradually increased. Christian Filipino settlers claimed lands and private holdings for fixed-plot commercial agriculture in the lowlands, dispossessing the Pala'wan of coastal lands and driving them further into the uplands. The Pala'wan were constructed as illegal squatters, since what remained of their ancestral lands had been reclassified as state timberlands. Centuries of state legal imperatives to expand and entrench private property rights, laws, and market value across the island's lowlands had been fulfilled, though they seldom penetrated far into the uplands. Soon, the unrelenting doctrines of evangelicals would build on colonial ideals, target Pala'wan uplanders, and attempt to reform them into God-fearing citizens.

The Biopolitics of Christianity and the Tribal "Other"

American colonial administrators were also obsessed with agentive objects (mutja, anting-anting) used by rural peoples for courage and power, often against colonial subordination (Ileto 1979, 22; Wheatly 2018). The supposedly animistic objects were infused with various supernatural and religious powers that blended Indigenous customs, rituals, and—in this case—Spanish-era Catholicism. Many messianic peasant groups used anting-anting to invoke the power of God, spirits, and other deities to render themselves invisible and invincible during the Filipino-American War (a.k.a. the Philippine Insurrection, 1899–1902). Some peasant groups, such as the Anting-Anting, were even named after the charm and feared by the US Constabulary and political elites for their tenacious and overwhelming resistance (Ileto 1979; McCoy 1982; Laurie 1989; Wheatly 2018). Although the use of anting-anting was not illegal, their associated beliefs and practices were denounced as primitive, animistic pseudoreligion. The US military connected their use to broader crimes against the colonial state and responded by organizing campaigns to locate and eliminate the objects and reform those using them (Wheatly 2018).

The secretary of the interior for the Insular Government, administrator Dean Worcester—a zoologist who deeply opposed independence—positioned the anting-anting and its users on the lowest rungs of the social evolutionary ladder. While, on the one hand, the anting-anting's alleged supernatural properties were dismissed as arbitrary and haphazard in use, on the other, the US administration feared that the use of "superstitious" objects would impede the adoption of the Christian faith and social advancement of their tribal subjects (Wheatly 2018, 6). The American colonial elite, whose religious and moral purity centered on Protestant ideals and convictions, believed that the sustained application of Protestant principles could discipline recalcitrant "natives" into abandoning supernatural objects and practices (2018). They distanced themselves from the Catholicism that Filipinos had inherited and fused with customary practices under Spanish occupation. Despite these reforms, however, most uplanders continued to use anting-anting in various aspects of upland life and livelihood. Much to the chagrin of missionaries, mutja remains a vital force among the Pala'wan today.

As the American colonial administration focused on cataloging and monitoring uplanders, broader population reforms strove to delineate and overcome the social and biological differences between lowland and upland populations. Military medical officers supported the war effort through comprehensive biomedical campaigns that attempted to insulate American soldiers some lowland populations from "fetid" tropical environments, foods, and cultures that caused tropical ailments and diseases (Anderson 2006). As the Philippine Insurrection dwindled, colonial officials and a growing number of military medics turned to "public health" campaigns embedded in broader scientific programs, infrastructure, and interventions aimed at reforming the bodies and souls of Indigenous peoples. Drawing on the racial ordering of the census, networks of hygiene inspectors and officers were dispersed throughout the countryside to educate lowland Indios and tribal uplanders about their unsanitary livelihoods and bodily practices. In the words of the chief health inspector of the Philippine Islands, Major Franklin Meacham, "countless unsanitary evils among the natives [would] be remedied," including slops, garbage, fecal accumulation, rubbish, and other debris (cited in Anderson 2006, 48–49). Although biomedical efforts were largely confined to the more densely populated lowland and midland settlements, US medical officers soon focused their biomedical efforts on upland populations. In subsequent decades, American colonial ideologies further incubated—or gave political license to—the motives of evangelical missions and environmental NGOs to venture further into the uplands to complete the state's incomplete biopolitical agenda.

Filling a Gap? Baptist Missionaries in the Philippines and Southern Uplands

For the American colonial government, Protestantism emerged as a potent instrument for wielding political influence across the predominantly Catholic archipelago (Raymond 2008, 48). There was no better tool than to allow ideologically aligned and often well-funded Protestant Baptist missions to fix their gaze on the hitherto untamed and uncivilized uplands–a nonstate space where the Catholic Church and American colonial rule had failed to pacify uplanders (Howell 2009). While peoples in the north of the southern islands of the country had adopted Catholicism to varying degrees, the predominantly Muslim lowland Moro peoples in the south and Indigenous uplanders (known later collectively as the Lumad on Mindanao, and Pala'wan in southern Palawan) retained customary beliefs or melded them with Islam (Eder 2010; Paredes 2013).[7] Evangelicals set out to change this reality.

Protestant beliefs fueled the American civilizing enterprise (Raymond 2008) by encouraging missionaries to pursue religious and nationalistic consolidation along the Philippine frontier (Paredes 2013, 38). They believed that "Philippine Catholicism [had] produced only superficial Christians . . . [with] little more than a veneer over a culture that remained, in important respects, heathenish" (Clymer 1980, 41). Driven by a desire to "penetrate" the Southeast Asian frontier, a steady stream of American volunteers soon supported Protestant missions, expanding their evangelical mandate into what were perceived as the tribal frontiers of the country (Clymer 1982; Raymond 2008).

By 1902, at least seven US-based Protestant missions led by Presbyterians, Methodists, Baptists, and Episcopalians, among others, had entered the country (Raymond 2008, 50). These missionaries, including the Seventh-day Adventists (SDAs), who arrived in 1905, were assigned specific territories to administer, and some did so on the basis of "comity," understanding that policies of cooperation and noninterference with other missions should be upheld (Clymer 1986, 32). The Adventists, however, went their own way. They neither joined the Evangelical Union nor aligned with any notion of comity respecting territorial divisions (Clymer 1986). By 1948, evangelical churches had spread throughout the northern, central, and southern Philippines; only Palawan was briefly spared from post–World War II evangelism, though this soon changed (Raymond 2008, 51).

Protestant evangelical missions adopted assertive biopolitical tactics of "supervising, punishing, conditioning, and correcting" Indigenous uplanders' customs and beliefs (Paredes 2006, 522). Unlike the Catholic pastors, who would "methodologically endear themselves to community" (Paredes 2013, 5), Protestant missionaries withdrew to focus on the transmission of religious doctrine through a range of strictures. While Catholic parishes and pastors often permitted the fusion of Catholic and Indigenous beliefs, the Protestants—and Baptists in particular—forbade Indigenous subjects from melding older beliefs and practices with their new faith (Paredes 2006). Baptist reformists retained Calvinist teachings of purification that separated out material and spiritual domains—including the use of objects such as anting-anting and the use of tingkep in ritual practices—to ascribe agency exclusively to the divine (Howell 2009). As Howell (2009, 258) notes, "the radical 'disenchanting' of the material world" was the "defining project of modernity" embedded in the religious work of missionaries across the Philippines. However, the Protestant missions went beyond changing Indigenous people's beliefs and practices; their goal was to comprehensively disrupt and purge Indigenous customs and worldviews (Paredes 2013, 49). This mission of reform in the seemingly "unreachable" non-Christian uplands was fully embodied by the SDAs in the Palawan frontier.

The Early Seventh-day Adventists

The SDAs emerged from Millerite evangelism in the United States during the 1830s (Bull and Lockhart 2007, 4). William Miller (and, later, Ellen White) spread Millerite preaching throughout New England to warn of the imminent "destruction of the world and [the second] coming of the Messiah" (4). When the world failed to end in the early 1840s, several members branched out to form a core group of so-called Adventists and reassessed aspects of their faith (Butler 1986). This core group had grown to about two hundred members in 1850; however, after the SDA church was formalized in 1863, the membership grew to three thousand (Butler 1986). Despite their attempts to bring their interpretation of scripture into the mainstream, the Millerite core of SDA beliefs was largely sustained.

It dictated that Sabbath be held on Saturday, and that disciples abandon practices that "polluted the soul," such as smoking, drinking, and eating meat or rich and spicy foods, so that they might ascend to Heaven, "the Sanctuary" (Butler 1986).

Adventists drew on the Old Testament to establish fundamental rules for the ideal Christian diet and behavior (Banta et al. 2018, 2). The consumption of certain meats was forbidden for fear it would "strengthen animal[-like] passions" associated with sexual desire, while consuming grains and fruits prepared one for "translation to Heaven" (Bull and Lockhart 2007, 90)—a healthy, divine diet consisted primarily of "seed-bearing plants" (Banta et al. 2018, 2). As I show later, Indigenous followers in the uplands were forbidden from harvesting, butchering, and eating "unclean" wild or domesticated animals, including omnivorous animals without split hooves (e.g., wild or domesticated pigs), riverine species without scales and fins (e.g., crabs, shrimps, and clams) and avifauna that eat fresh meat or carrion (e.g., owls, bats, storks, herons, vultures, and seabirds). Meat from such "unclean" animals would supposedly pollute body and soul and impede divine healing. These meats, however, have long served as crucial sources of protein in the otherwise protein-poor, carbohydrate-rich diets of Indigenous highlanders (Banta et al. 2018).[8]

The SDAs wasted no time in dispersing American and Australian evangelists to the Philippines in 1905 and 1906 to convert, baptize, and build new churches in Manila (Land 2015). The country's first Adventist church was established in 1911 and consisted of twelve members, six Filipino converts and six American missionaries. The Filipino converts were critical to the SDA's early mission: well-versed in Indigenous languages and cultures, they built the first Indigenous church to recruit additional members from the local communities surrounding Manila (Fernandez 1990). Several of the original missionaries also returned to Australia in 1912, bringing "at least five more missionaries with them" to evangelize beyond the Tagalog regions of the country (136).[9]

After opening the Philippine SDA Academy in Manila in 1917, evangelical missions proceeded to northern Luzon and other remote regions to convert Indigenous uplanders well into the independence era. Most SDA infrastructure remained headquartered near Manila, including the newly established Adventist University of the Philippines; however, in the 1950s, a new SDA sanatorium and hospitals were built in outer regions, such as Cebu City and Davao City. By the 1960s, multiple evangelical centers had been established for the purpose of training Filipinos, and God's spoken word was now broadcast in several languages over SDA radio waves throughout the country (Land 2015). The Philippine SDA membership base increased from 20,000 in 1939 to 70,000 members by 1962, with a growing rural presence (2015).

SDA and the Adventist Frontier Missions

From the 1960s, many SDA institutes, seminaries, and graduate colleges began training Filipinos in religion, health, nursing, and other biopolitical vocations in the country's remoter regions. Given that becoming a medical missionary promoted upward social mobility in the faith (Land 2015, 14), an expanding network of SDA missions soon established elaborate self-sustaining camps with hospital services, living quarters, and churches in upland frontier areas. The SDA soon became known as the "medical missionary people of the world" (Bull and Lockhart 2007, 14).

The Adventist Frontier Missions (AFM)—founded in the United States—was considered the most venturesome and zealous in its mission to extend God's "reach to the unreached." Aligned with the main SDA Church and its beliefs, the self-financing AFM worked with a degree of independence, affording it flexibility to share the Gospel of Jesus Christ among those heathen "tribes" deemed the most primitive and savage, and most in need of direction and support.[10] By 1987, the AFM had sent its first missionary family from the United States to "plant the church among the Ifugao people in northern Luzon, Philippines" (AFM 2023). Six years later, one hundred Ifugao had been baptized. Approximately eight years later, the SDA following had replicated itself across the region (2023), including in Kamantian, southern Palawan.

As the SDA and AFM gained traction in the southern uplands, the state's forest governance architecture shifted from centralized command-and-control approaches to decentralized reform initiatives in the late 1990s and early 2000s. These governance reforms granted nonstate actors greater political leverage to work independently or to govern the ungoverned more closely, on the state's behalf. Two types of evangelical missions with overlapping objectives—religious organizations and environmental NGOs—would soon discipline Indigenous bodies and souls in upland spaces through sustained biopolitical reforms.

Postcolonial Biopolitical Continuities

In the postcolonial period, the Philippine Constitution and land laws maintained the essence of colonial property rights law and forest governance policies to comprehensively criminalize uplander worlds. Ferdinand Marcos's dictatorial regime (1961–85) further centralized forest and land use governance to curb swidden and expand commercial agriculture (Vitug 2000; Dressler 2009). Under Marcos, the Philippine government enacted multiple legal and policy provisions that were designed to eradicate swidden cultivation and protect timber stands across the country. In 1963, the revised kaingin (swidden) law imposed additional penalties on "kaingineros [sic]" (Scott 1979); in 1965, a National Conference on the Kaingin Problem sought to identify and resettle swidden cultivators (Population Center Foundation 1980, 11); and, in 1971, the Kaingin Management and Land Settlement Regulation aimed to integrate uplanders into state conservation programs so as to curb swidden encroachment into state forestlands (Population Center Foundation 1980).

In 1975, Marcos's Presidential Decree (PD) No. 705 (The Revised Forestry Code) served to classify, manage, and utilize forests in the public domain. Reinforcing the upland and lowland divides, the law specified that all lands with a minimum slope of 18 percent and areas above six hundred meters belonged to the public domain or timberlands. Although most Indigenous uplanders had long occupied upland forests and considered these ancestral lands necessary for survival, state foresters maintained that mountain slopes, valleys, and ridges were illegally occupied or even completely vacant. The law laid the groundwork for forest land occupancy programs designed to facilitate state and, later, nonstate monitoring and control of uplander activities and incentivize agroforestry, afforestation, and sedentary ways of life.[11]

During the Marcos regime, a suite of tenurial programs was rolled out across the country with the aim of optimizing uplander livelihoods. These included Communal Tree Farming (CTF) in 1978, the Family Approach to Reforestation (FAR) in 1979, and the Integrated Social Forestry Program (ISFP) in 1982, among others. These community agroforestry initiatives often followed similar experimental designs that tried—and typically failed—to sedentarize and improve uplander livelihoods. The standard approach was to offer uplanders twenty-five-year leases with various conditionalities, including tree planting and monitoring rather than slashing and burning. Such institutional designs would supposedly encourage people to abandon swidden and adopt intensive fallow management in a single territory. Noncompliant activities, such as clearing older forests, resulted in the revocation of tenurial leases. Most decentralized upland programs did more to regulate swidden than offer livelihood support.[12] Soon, however, various nonstate actors emerged as expert brokers, mediating the use of state tenurial policy to provide uplanders with a sense of citizenship and security, albeit temporary and conditional.

A New Dawn? Or the Same Old Biopolitical Song?

After a popular uprising in the Philippines, known as the People Power Revolution, overthrew the Marcos regime in 1986, incoming President Cory Aquino's constitutional reforms led to renewed democratic vigor and a flourishing of Indigenous rights and environmental NGOs.[13] Many of the underground social movements that had resisted elite political control and the exploitation of human rights, lands, and forest resources were now free to mobilize and outwardly support Indigenous rights and environmental agendas (Clarke 1998). NGOs could fully exercise their moral desires and duties of assisting the poor through local reform initiatives, now backed by liberal state laws.

Article 12, Section 5 of the amended Philippine Constitution (1987) formally acknowledged Indigenous rights to land and recognized civil society as a legitimate partner in state governance. It was during this period that many underfunded state agencies turned to increasingly well-funded and organized NGOs and bilateral organizations to implement the Indigenous rights agenda nationwide (Eder and McKenna 2004). In time, new NGO-state partnerships emerged, leading to a range of NGOs adopting de facto state roles and responsibilities in remote upland and coastal areas. However, despite the new constitutional mandate to "protect the rights of Indigenous cultural communities to their ancestral lands and to ensure their economic, social, and cultural well-being," securing de jure land rights for Indigenous uplanders proved difficult (Constitution of the Republic of the Philippines 1987). State hesitancy to release lands under Indigenous people's control soon became evident in the complex legal and bureaucratic constraints underpinning full recognition of Indigenous ownership and management of ancestral lands. The Regalian Doctrine was alive and well (Dressler 2009; Theriault 2019).

After the Aquino government established the Commission on Indigenous Cultural Communities and Ancestral Domain in the late 1980s, the US Agency for International Development (USAID) invested $125 million in the state-managed Natural Resource Management Program (NRMP) in 1991 (McDermott 2000). The NRMP was Southeast Asia's first major foreign-funded state intervention to involve Indigenous land claims. With this funding, the Department of Environment and Natural Resources (DENR)—now working closely with bilaterals and NGOs and staffed by former activists—drafted and issued the Departmental Administrative Order No. 2 (DAO 2, 1993). The DAO 2 established guidelines for implementing so-called Certificates of Ancestral Domain (or Land) Claims, or CADCs (McDermott, 2000; Dressler, 2009; Theriault, 2019). NGOs, such as the paralegal Indigenous rights group PANLIPI (Tanggapang Panligal ng Katutubong Pilipino), directed the content of DAO No. 2 and other national Indigenous rights policies. As a staff member of the Palawan-based Environmental Legal Assistance Center (ELAC) explained in a 2001 interview: "It was the PANLIPI lawyers that developed the content of the Administrative Order. I used to be a PANLIPI lawyer in 1992 until 1993. . . . The DAO [No. 2] gave us the opportunity to have Indigenous people's ancestral claims and livelihood needs, such as rattan, at national parks because they were getting short-changed by the [migrant] communities there."

DAO No. 2 was the first national policy that sought to define and regulate how ancestral domains in the Philippines were identified and delineated. A new Provincial Special Task Force on Ancestral Domains was soon established to initiate national consultations through so-called local cultural community offices and NGOs, which (then as now) served as essential local-state brokers for the delineation and management of ancestral domains in upland areas. In the first few years, several CADCs were established on southern and central Palawan Island with the aim of granting Indigenous communities collective ownership over, access to, and use of ancestral lands, including forests and nontimber forest products (NTFPs) (McDermott 2000; Dressler 2009; Theriault 2019). It was the motivated staff of NGOs who proved crucial in facilitating the consultations, designs, and implementation of CADCs in remote upland settings that many state officials avoided.

In 1997, the Philippine Congress finally ratified the landmark Indigenous Peoples' Rights Act (IPRA, 1997), which afforded Indigenous peoples de jure land rights under Certificates of Ancestral Land and Domain Title (CA[L]DTs; see Theriault 2019). Although politically bold and necessary for the time, IPRA (1997) and DAO No. 2 were ultimately informed by narrow, essentialist notions of indigeneity drawn from the colonial era. The discursive substance of the legislation and its rollout drew on reified cultural categories and sentiments about Indigenous peoples' livelihood practices, residence, and customary practices long held by the state and increasingly NGOs.

The legislation allowed Indigenous communities to claim larger CADTs, provided that their customary characteristics were sufficiently traditional and sustainable. For the state body, the National Commission of Indigenous Peoples (NCIP), only sufficiently customary and homogenous "Indigenous cultural communities" living in a predefined ancestral territory could claim and occupy ancestral domain titles. Communities hoping to qualify for a CADT had to demonstrate continuous occupation and land use, along with cultural traits that rendered them "tribally distinct" from other non-Indigenous religions, cultures, and most lowland Filipinos (IPRA 1997, Chap. 2, Sec. 3h).[14] By extension, Indigenous peoples holding CADTs had to uphold sustainability conditions, such as "maintaining ecological balance," "restoring denuded areas," and "observing laws" (1997). Swidden was largely denounced, and harvesting commercial timber was prohibited. In this way, colonial and postcolonial tropes of the tribal Other had manifest in the logic and practice of securing Indigenous rights and forest conservation among both state and nonstate actors in the Philippines. By the late

1990s, as state and nonstate partnerships weakened, more flexible and mobile NGOs assumed greater responsibility for implementing ancestral domains and livelihood programs through an ambiguous mix of Indigenous recognition and conservation conditionalities.

The IPRA's rigid conception of indigeneity and notions of ecological balance demanded that Indigenous peoples align their complex, fluid, and changing customs and livelihoods with rigid and absolute management criteria as prescribed in Ancestral Domain Sustainable Development and Protection Plans (ADSDPPs). As state brokers and quasi-independent actors, these NGOs were now frequently visiting, and often embedded within, remote rural communities, drawing on ADSDPP criteria to impose biopolitical ideals on the lives and livelihoods of upland peoples. They assisted in surveying cadastral boundaries, completing timber and nontimber resource inventories, drafting management plans with insertions of ecological knowledge and rigid kinship diagrams, and devising land use histories and other items that assisted in delineating, filing, and managing CADTs. NGOs increasingly steered and influenced community organization, hierarchy, and identity politics in line with reified cultural categories and state-imposed governance structures not entirely dissimilar from the racialized census of the US colonial era (e.g., tribal chieftains and tribal councils) (Theriault 2019).

As Hirtz (2003) notes, such Indigenous policy procedures emerged for the twin purposes of enabling the right of recognition (the right to be an Indigenous people or group) and imposing specific obligations, ideals, and conditions upon Indigenous existence (the right to act, in a certain way, as an Indigenous people). Recognizing indigeneity in law and policy, Hirtz argues, emerged as a discursive political expression of late liberal modernity. The sustained production of tribal and traditional characteristics was set against, legitimated, and formally instituted as a remnant artifact of precolonial times subject to modern rights and conditions (e.g., of sustainability and cadastral property rights; see also Hobsbawm and Ranger 1983). For some, state and nonstate actors defined a notional indigeneity that was deliberately connected to "a suitable historical past," which had to be reformed to align with a suitable modern future (Hobsbawm and Ranger 1983, 1).[15]

In time, an increasing number of NGOs and people's organizations adopted state-like roles and responsibilities in remote upland areas. NGOs, with greater trust and authority to support Indigenous peoples, took on diverse portfolios of de facto state duties (Hilhorst 2000). They considered themselves closest to the people, aware of local needs and concerns, and well placed to broker rural support programs, often on behalf of state agencies (Bryant 2005). Indigenous-rights, environmental, and even missionary entities found themselves mediating state-society relations in frontier settings, maintaining moral ground by offering promises and hope to poor uplanders (Fisher 1997).

Conflating essentialized notions of the tribal Other and indigeneity more broadly, NGOs invoked a rhetoric of hope that interventions would generate optimal outcomes through "tenurial security," "agroforestry solutions," and, increasingly, biopolitical provisions such as health care, clean water sources, and primary evangelical education (Dressler et al. 2010; Dressler 2017). Across the uplands, nonstate actors worked to deliver a collective, morally good, and promised future, one that was discursively reproduced and leveraged through their own political objectives, but often contested by Indigenous peoples with contrasting livelihood priorities and political aspirations (Mitlin et al. 2007).

In the early 2000s, Indigenous rights and environmental NGOs experienced a decline in funding as donors reoriented their priorities; meanwhile, missionaries sustained their financial base and ideological reach further into the uplands.[16] Conservation NGOs, in particular, attempted to add market value to those customary resource uses that were deemed less extensive and extractive (e.g., basket weaving and honey gathering) in the hope of curbing their exploitation and that of other forest resources, as well as eradicate those activities considered more extensive, extractive and degrading to forest environments (e.g., swidden). For many NGOs, the overriding imperative soon became "selling nature to save it" (McAfee 1999), which, in practical terms, meant leveraging both punitive strictures and the assumed economic value of species (or ecosystem services) to incentivize "alternative uses" and "stewardship" among highlanders (Fletcher et al. 2019). Among Baptist missionaries, the biopolitical imperative was even deeper and more unrelenting. As the next chapters show, through complex infrastructure and practices they aimed to incrementally and powerfully reform Indigenous social relations, modes of reproduction, customary beliefs, and livelihoods to align with purity, docility, Jesus, heaven, and hell. These typically unsolicited biopolitical reforms were neither spatially isolated nor ideologically discrete; myriad projects aiming to reform Indigenous livelihoods and ways of life increasingly converged in the same upland spaces.

The American colonial regime built on Spanish racialized conceptions of uplanders to justify a raft of laws and measures that constrained Indigenous lifeways and livelihoods to facilitate incremental control of seemingly distant upland peoples and forest landscapes. The US administration encouraged the spread and influence of Protestant beliefs and partly opened the door for Baptist missionaries to deepen proselytization in highland areas where the Catholic Church had failed to fulfill biopolitical reforms. As I show next, Baptist missionaries worked with remarkable patience and persistence, sustaining their efforts to reform uplanders into pliant Christian subjects well into the postindependence era. While the democratically elected Aquino government oversaw decentralization and a rise of Indigenous rights in the 1980s, these reflected narrow and reified conceptions of indigeneity that—like Baptist missionaries—aligned with biopolitical practices and instruments (i.e., fixed tenure and sedentarism). Lingering from the earliest days of Spanish rule, such comprehensive Bureaucratic Orientalism (Hirtz 2003) has saturated the imaginaries of civil society across the country. It upholds the idea that Indigenous peoples are vulnerable, customary peoples that require exceptional interventions and assistance that optimize in line with dominant societal ideals.

Influenced as they were by the country's colonial ideology, legislative legacies, and racialized political economies, environmental NGOs and Adventist missions both fixed their gaze on the uplands with overlapping agendas. They set out to locate and reform those Indigenous peoples who apparently indiscriminately slashed and burned forests and clung tenaciously to customary ways of life. Success was measured in incremental reforms—that is, by how many of these "forest destroyers" were sedentarized or converted into Christian disciples. Their ultimate goal was the complete reworking of Pala'wan lives and livelihoods.

As Li (2014, 13) notes, "Frontiers are not only characterized by lack. They are simultaneously coveted places, envisaged by various actors as sites of potential." In places that lay beyond the state's reach, nonstate actors undertook biopolitical reforms to instill new "imaginings, expectations and visions" of such potential (Borup et al. 2006, 285–86). Sustained reforms affected both family and village life through a range of lingering promises, incentives, and strictures. Nonstate programs and projects thus aimed to govern by gradually reforming how uplanders assigned meaning to and invested in social relations, custom, and livelihood practices.

The next chapter shows how civil society interventions have become more prevalent, efficient, and effective than those of state agencies in the uplands. Well-funded nonstate organizations of various types—from SDA missions to international NGOs, such as Conservation International (CI)—have had a particularly enduring presence in southern Palawan. Each organization's moral imperatives and practices overlap discursively and materially in ways that gradually reform upland peoples, livelihoods, and ancestral territories. I describe how Pala'wan worlds are being reformed through the conjuncture of historical political legacies and contemporary political, economic, and ecological changes within and beyond the uplands. I show how certain sociomaterial objects, particularly the tingkep, came to mediate rural and urban social relations and livelihoods across diverse rural and urban geographies (Massey 1993, 67). I elaborate on how Indigenous territories are being reframed and transformed into nonstate biopolitical spaces: areas of experimentation and sustained reform.

OF FORESTS AND GODS

The Baptists are in the interior, the Catholics are in the lowlands.
I know this because I was affiliated with them [the Baptists] for five
years. My father was sad when he found out, but since he was a
Catholic priest himself, he did not question me. He actually convened
my marriage at the time. But I went back to being Catholic because
they are looser; those in the interior [SDA] stopped us from living
our culture. They restricted our use of traditional herbals. . . . Such
things are prohibited, and it causes other IPs (Indigenous peoples) to
further doubt their older traditions.

—Panglima Milandro Tiblack

In southern Palawan today, the lives of uplanders are beset by an ambiguous mix
of opportunities and constraints that emerge from the intersection of decline,
reform, and optimization. Converging political and ecological processes, both
historical and contemporary, influence the origins and rise of nonstate practices of reform and the racialized logics determining which lives and lifeways
are useful and worth enhancing, and which are threatening and warrant erasure.
This chapter describes how such nonstate biopolitics manifests in the uplands
of Palawan and considers who was (and was not) "authorized to take part in
debates and decisions about what life should be protected, how, why, and for
whom" (Biermann and Anderson 2017, 1). I trace the recent biopolitical conjunctures that have enabled the rise of environmental NGOs and Adventist Frontier missionaries—heterogenous entities with intersecting motivations to reform
Pala'wan social relations and livelihoods. I analyze the differential impacts and
outcomes of these socioecological reforms among forest-reliant Indigenous
uplanders as they negotiate uneven agrarian political economies.

In what follows, I describe four intersecting social histories: 1) the origins
and practices of grassroots NGOs on Palawan; 2) the rise of parastatal enti-
ties (quasi-NGOs) such as the Palawan Tropical Forest Protection Programme
(PTFPP); 3) the market-based turn and architecture of the environmental NGO,
Conservation International (CI); and 4) how the Adventist Frontier Missions
(AFM) profoundly deepened biopolitical reforms in the highlands of southern
Palawan. The chapter details how the parastatal PTFPP "governed the way" for

environmental nonstate actors to subjugate and discipline the bodies, behaviors, and livelihoods of the Pala'wan in line with market ideals and incentives. Crucially, this included valorizing the tingkep and other customary objects as "sales objects" (i.e., commodities) to further incentivize and induce reforms aimed at curbing swidden-based livelihoods. In the process, as Ferguson (1994, xiv) noted decades ago, NGOs such as CI have developed their own discursive logic that constructs upland entities and issues as particular objects of knowledge and management concern—that is, an "anti-politics" of governing the tribal Other, where degrading swiddens become a matter of technical intervention, despite their inherently political nature. Nonstate actors have done remarkably well in reifying, simplifying, and misrepresenting Indigenous ways of life in ways that are legitimated, unquestioned, and acted upon accordingly.

I draw on program documents and my own fieldwork to describe how these nonstate actors' programs and projects leveraged uplander identity and customs to induce converging biopolitical logics and interventions among the Pala'wan. Understanding this recent biopolitical history sheds light on how environmental nonstate actors have used similar discourse and local brokers to "double down" in the uplands today—the topic of chapters 6 and 7. As I show, they variously deployed hygienic bodily practices, environmental education and stewardship, and alternative livelihood projects to encourage the "stepping out" of swidden and "stepping into" sedentary agriculture. In various ways, the tingkep mediated all these practices of reform.

I then consider the AFM's rise, structure, and interventions as well as their influence on ways of life in southern Palawan. The AFM aimed to reform Pala'wan social reproduction, identities, custom, and livelihoods—the same characteristics that nonstate conservationists harnessed to optimize Pala'wan existence to protect forests and biodiversity. I show how Pala'wan social relations and customary practices, such as tingkep making, have become the subjects of intensifying reforms at key biopolitical conjunctures involving both sets of nonstate actors. Environmental NGOs integrate market-based schemes and strictures that valorize and sanction certain Pala'wan customs and material culture, while criminalizing swidden and hunting, to facilitate livelihood change for long-term forest conservation. Meanwhile, the AFM prohibits myths, rituals, and certain livelihoods in the interest of expunging Satan and bringing the Pala'wan closer to God. These multiple practices of reform converge on the tingkep, its makers, and nonhuman worlds. Nonstate actors either invoke or suppress Pala'wan indigeneity and associated materiality to create an "instrumental imaginary," a "crucial marker for appropriate ways of living" in upland settings (Chandler and Reid 2019, 1).

I describe how both environmental NGOs and missionaries' broader agendas, infrastructure, and practices aimed to enroll Indigenous uplanders into specific social contracts and agendas of reform. Despite being actively contested, the social contracts between certain Pala'wan and nonstate actors continued to be rolled out to provide incentives, services, and opportunities in exchange for cooperation and alignment with nonstate aims and objectives (Simpson & Pellegrini 2022; Titeca et al. 2020). Such contracts, I argue, aimed to reproduce a biopolitical architecture with degrees of permanence and stability for upland peoples in the state's absence, despite the despondency of many toward such agreements (Simpson and Pellegrini 2022).

Resurgent Biopolitics in Palawan

Palawan's frontier status, immense natural wealth, and aesthetic beauty have drawn corresponding struggles to preserve, appropriate, and control its forest resources (Eder and Evangelista 2015). As in previous decades, the 1990s onward saw an increase in the number of Christian and Muslim migrants leaving the country's conflict-ridden areas (e.g., from Luzon, the Visayas, and Mindanao) to settle the island's central and southern lowlands.[1] The island's population grew from 317,782 in 1980 to 528,287 in 1990, a growth rate of 3.58 percent (Eder 1999), with 1.2 million people living there in 2021 (PBS 2022). However, far fewer migrants have settled in the island's sparsely populated, mountainous southern region (Dressler 2009), which still retains extensive forest cover. Here, Pala'wan uplanders remain far less integrated in lowland political economies, state institutions, and Christian denominations than elsewhere (Eder and Fernandez 1996).[2]

In the 1990s post-Marcos era, international and domestic NGOs adopted broad mandates that aimed to curb forest exploitation and champion Indigenous rights to land and livelihoods on the island (Vitug 1993). The new political openings emerging under the Aquino (1986–98) and Ramos governments (1992–98) allowed nonstate actors to push for national and subnational legal reform agendas (Dressler 2009) for conservation and Indigenous rights (e.g., DAO, 1992; IPRA, 1997; and the National Integrated Protected Areas Act, 1992). Many lawyers-cum-activists drew on their law school days and activist networks at the University of the Philippines Diliman and Los Baños to form grassroots NGOs and people's organizations (POs) with friends and colleagues. In Puerto Princesa City, most organizations depended on meager funding and volunteer labor to survive (Dressler 2021).

Many nonstate groups expanded their programs across rural settings in central and, later, southern Palawan through local brokers and activists. Such groups included the Indigenous Peoples' Apostolate (IPA, now the Commission on Social and Special Concerns, or CSSC); the United Tribes of Palawan (Nagkakaisang mga Tribu ng Palawan, or NATRIPAL); the Legal Assistance Center for Indigenous Filipinos (Tanggapang Panligal ng Katutubong Pilipino, or PANLIPI); the Environmental Legal Assistance Centre (ELAC); and the umbrella group the Palawan NGO Network Incorporated (PNNI).[3] These NGOs received an influx of foreign and state funding to establish community-based forest management initiatives within Pala'wan, Tagbanua, and Batak ancestral domain claims and titles. Programs and project funds flowed to NGOs on the assumption that they were closer to forest dwellers and better positioned to garner trust and establish community conservation schemes informed by "traditional ecological knowledge" and "Indigenous management systems." Most of these NGO-driven community-based projects were part of state programs—a hangover from Marcos's forest policy architecture—concerned with valorizing and implementing tenurial provisions, agroforestry, and fixed-plot commercial agriculture (e.g., intensive wet rice cultivation) to eradicate swidden. PNNI was established in 1989 to help coordinate and allocate such projects, as the former director noted:

> PNNI started at a time when many development projects were coming here, and no one was examining whether they were suitable for the province. It was all being done through a top-down approach. But the many NGOs that started up did not coordinate with each other; they attempted to operate on their own.
>
> So the local Indigenous rights advocates and environmentalists sat down together and talked about how they could stop this from happening. We said these development projects could not be done in Palawan without setting up critical collaboration and consultation whereby the local folks could have something to say about what is being brought into Palawan.

Building on their records and reputations (Hilhorst 2000), these NGOs and rural activists drew on influential church, PO, and village social networks and projects to facilitate livelihood support programs (Pinto 1999) that aimed to incentivize and optimize resource production and income generation at the household and community level. The biopolitics of modern subject-making was thriving.

Unlike big international environmental NGOs such as CI and WWF, grassroots NGOs regarded Indigenous peoples as historical political allies in social movements and project delivery, often prioritizing Indigenous needs, concerns, and rights over their own project mandates (Dressler 2021). This was particularly true for several Indigenous rights–focused ecumenical NGOs: under the leadership of a Tagbanua priest, the Catholic Commission on Social and Special Concerns (CSSC, under the Apostolic Vicariate of Puerto Princesa) established rural POs to delineate the first ancestral domain claims in central and southern Palawan. In contrast to coercive evangelical missions, the CSSC's religious services, education, and livelihood initiatives allowed Indigenous peoples to decide for themselves how best to integrate Catholicism with their own beliefs. In upland settings, such situated interfaith activism was relatively receptive to and tolerant of other Indigenous customs.[4] Working with NATRIPAL and other grassroots NGOs, the CSSC pioneered the capacity-building and empowerment of Indigenous communities under the watchful "eyes of God" but with some space for customary beliefs and practices.

Despite the productive alignments between some grassroots NGOs and Indigenous peoples, many local NGOs and Philippine chapters of larger NGOs continued to draw on and engage Indigenous peoples based on essentialized assumptions and representations of the uplanders they were meant to partner with and offer political support. For environmental NGOs, the recent recognition of Indigenous rights to land and forests placed greater responsibilities for forest stewardship and sustainable management on uplanders. Most NGOs soon qualified and tempered their support for upland livelihood practices—particularly swidden agriculture—that might eventually lead to "degrading" impacts and outcomes. During an interview I conducted in 2001, a staff member of a local environmental NGO lamented that, once modern ways corrupted sustainable Indigenous swidden cultures (e.g., "ecological long fallow"), they might turn to destructive slashing and burning (kaingin) and therefore warrant closer monitoring and management:

> From our perspective, we agree that the kaingin method should be part of management plans. So we believe that the slash and burn method of the katutubo are compatible, but the compatibility of the slash and burn of twenty years ago is not at the same level today [2001]. Right now, the situation is different. With the ten thousand hectares of CADC, there is much forest destruction already. The forest is threatened. But a fallow period of even ten years is not even enough. Currently, it is three to five years, that's not enough, even five years is insufficient. I would say that the cultural practice of slash and burn has metamorphosed. The metamorphosis is not entirely voluntary, but there is an external factor pushing it because of the threats. We've got migration and destruction. The state of the environment here in Palawan is such that you cannot just allow slash and burn wherever as it used to be.

Such biopolitical ideals are almost inevitable as respect for uplander rights become conditional. Many NGOs explicitly promote Indigenous rights for as long as they can leverage the politics of indigeneity for rural mobilization, campaign building, or accessing international funds. As soon as hints of localized forest degradation emerge, Indigenous livelihoods are managed and fastened to "tribal tradition." Bureaucratic Orientalism sets in (Hirtz 2003).

During and after Marcos's rule, the south's stigma of remoteness, hosting "dangerous Tausug," and deadly forest diseases partly spared the region from major logging and mining operations, deforestation, and violence observed elsewhere in the country. NGO activism was also (later) instrumental in curbing the expansion of extractive development. Informed by colonial tropes of indigeneity, many nonstate actors continued to reify and leverage Palawan's tropical forests and Indigenous customs as powerful emblems for both protection and reform. Such strategic essentialisms allowed the NGO Haribon Palawan and others to successfully petition for an island-wide moratorium on logging that was supported by tourism operators, Department of Environment and Natural Resources (DENR) employees, and other activists.[5] Early on, the biopolitics of sustainability required Indigenous peoples to have—or at least to convey—intact traditional beliefs, ecological knowledge, and management systems seemingly necessary to conserve upland forest landscapes.

The Palawan Tropical Forest Protection Programme

Amid the rising tide of global environmentalism, UNESCO declared Palawan a biosphere reserve in 1990, carving up the island with regimented management zones under the so-called Palawan Integrated Development Project (PIADP), a well-resourced, comprehensive parastate biopolitical program funded by the European Economic Community and the Asian Development Bank (ADB). The PIADP (1982–90) aimed to economically develop Indigenous communities and arrest environmental degradation in Palawan through "forest stabilisation" projects (ADB 1991, 4). The main project's main components centered on optimizing Indigenous uplanders' existence through productivist interventions including irrigation schemes, livestock development, agricultural intensification, and diversification initiatives. A central program pillar involved the ambitious, landscape-scale "upland stabilization program" complete with agroforestry interventions and land tenure certificates (e.g., Integrated Social Forestry schemes) that were intended to transform upland swidden farmers into settled, fixed-plot agriculturalists who maximized crop yields (4). These ambitious management objectives would inform upland zoning and project interventions for decades to come.

In 1992, the PIADP's groundwork produced the Palawan-wide Strategic Environmental Plan (SEP) and its own bureaucratic creature: the EU-funded Palawan Council for Sustainable Development (PCSD). The PCSD retained broad authority to manage the island's natural resources while initially accommodating NGO involvement in decision-making and governance. The SEP launched the Environmentally Critical Areas Network, or ECAN (PCSD 1992), a graded zoning scheme that divided the island into sequenced resource use areas with increasing restrictions on activities along the coastal-mountain gradient, culminating in a "no-touch" core zone at higher elevations (1992). Despite being difficult to enforce, this zoning system emerged as the PTFPP's operating theater for rendering upland peoples and landscapes legible, manageable, and modern.

The PTFPP was launched in 1995 by the PCSD as a seven-year upland development and conservation entity. It served as a parastatal institution with the versatility, reach, and capacity of an NGO. Its governance aim was to support the SEP and ECAN zoning by implementing forest conservation and livelihood development initiatives in the uplands of Palawan. Interventions unfolded in both traditional (at 100 meters above sea level [m asl]) and controlled-use zones (100–500 m asl) to prevent swidden farming and nontimber forest product (NTFP) harvesting from encroaching on the buffer zones' restricted use areas (500–1,000 m asl) and the older forests of the exclusive core zone (1,000 m asl). Alternative livelihood initiatives and the provision of social services—typically the responsibility of central governments—were implemented in the traditional-use zone. A range of integrated interventions—agroforestry, contour farming, nursery production, livestock rearing, informal education, health services, anti-burning initiatives, and environmental awareness campaigns—were designed to bind Pala'wan households and livelihoods to mid-to-lower elevations. These locations were sufficiently distant from the older forests of the core zones and sufficiently close to lowland areas, ensuring more efficient sedentarization and integration of families into the lowland political economy (PTFPP 2000). In this way, the PTFPP laid the governance architecture for other nonstate actors to subsequently pursue their own reform programs.

From 1995 to 2002, PTFPP staff worked across eleven major catchment areas from central to southern Palawan. In the Mount Mantalingahan Range, the Inogbong and Tamlang catchment areas hosted the main, longer-term PTFPP projects (Marenshewan sits just south of the Inogbong project site, and Kamantian sits in the middle of the Tamlang catchment management region). In both catchment areas, technical project managers, community organizers, and enforcement officers attempted to stabilize swidden fallows in midland areas above one hundred meters and reforest areas above five hundred meters in upper elevations. The PTFPP attempted to "lock" swiddens into multiple- and traditional-use areas to protect older growth forests in higher elevations from rotational clearing and burning. Project officers labored tirelessly to convert messy swidden fallows into economically enriched fallows, broadly equivalent to agroforestry (Dressler 2014). In the Inogbong catchment, the PTFPP and its staff strove "to ensure that sustainable land uses are provided with enhanced income opportunities for Marginalized Upland Groups and IPs (Indigenous peoples) that would decrease extraction pressure on the remaining natural resources" (PTFPP 2002b, 28). The idea was that "alternative livelihood" initiatives would reduce forest resource dependency without the need for any resettlement (Smith and Dressler 2018).

Although project officers wanted trees in the ground, they were not the ones planting them. Mostly, PTFPP staff enrolled respected customary leaders (panglima) as their cultural brokers and ecological laborers under the awkward (though revealing) title of "upland development leaders." As development leaders, their job was to exert influence on village-level decision-making to ensure enrollment in, and compliance with, forest regeneration and swidden stabilization interventions (Hobbes, 2000, 53). Several panglima brokered program deliveries on behalf of project staff across upland villages, encouraging families to support the short- and long-term livelihood outcomes of the projects.

The panglima also frequently liaised with PTFPP staff to learn more about project terms of reference and outcomes. Just as panglima reported on the social and material progress of the upland development interventions, they and other Pala'wan also attended seminars clarifying the rules of ECAN zoning and the SEP law. The seminars explained, in no uncertain terms, that clearing and burning "new" forests for swidden should be avoided, that it was strictly prohibited to do so in the older forests in core zones, and that any infractions would undermine "development" opportunities. As an accessory, the meetings included personality coaching to optimize leadership confidence, conflict management skills, and community facilitation and awareness skills as part of established agreements with communities. Such leadership hierarchies and alignments—more reflective of lowland barangays than their own social relations—were imposed and extended to incrementally discipline fellow farmers in the uplands (Macdonald 2007).

Project panglima were assigned upland jurisdictions in which to convey different rules, regulations, and responsibilities. The PTFPP and lowland politicians in the Inogbong area began by drafting specific rules relating to controlled burns for panglima to communicate and enforce in their upland community. The new rules governing swidden burning dictated that a 1.5–3 meter firebreak (gahit) must be cut around burning fields, that wind strength and direction must be considered, and that the entire community be present to suppress any "escape fires." Ironically, most Indigenous peoples on Palawan have long managed swidden fires using these methods (Dressler et al. 2021). Despite Pala'wan fire knowledge, the strictures condescendingly advised them to refrain from clearing and burning forests for swidden (Hobbes 2000, 55; PTFPP 2000).

The PTFPP also broadcast biopolitics over the local airwaves. It established its own radio show to transmit information about the program's objectives and the appropriate uses of forest landscapes further into the uplands. By distributing radios to upland households as a reward for planting and maintaining two hundred tree seedlings and distributing trees to other families, project staff ensured that their ecological overtures resonated within the homes and villages of the Pala'wan (Hobbes 2000). The PTFPP's directives were broadcast by gifting radios already tuned to *Oras ng mga Katatubo* (The Indigenous peoples' hour), a program featuring panglima reiterating project doctrine and the value of planting trees for income and reforestation (in "degraded" swidden areas). Some panglima even supplemented the broadcasts with field visits "to monitor new [contract] rules and tree planting activities on request of the programme" (Hobbes, 2000, 65). Respected Pala'wan leaders emerged as quasi-disciplining agents.

For the many Pala'wan families still dedicated to swidden, fallow enrichment by planting "higher value" tree crops made them less inclined to clear and burn fallows in subsequent years. These lands were now permanently taken out of rotation. The intensive managing of short fallow plots led to diminished soil fertility and declining yields, causing farmers to seek out more fertile land in older forests by breaching ECAN zoning (Dressler 2009, 2013; Smith 2015). While enriching fallows with commercial tree crops might provide farmers with a limited income decades later, the intervention curbed the rotational cycle and associated socioecological practices of swidden cultivation (e.g., sharing, subsistence yields, ritual) among the Pala'wan. Ultimately, the predefined zoning systems, associated rules-in-use, and varied incentives centered on transforming upland farmers into modern subjects living in commodified upland landscapes.[6]

The Tamlang Catchment: Incentives and Value-Added Production

In 1992, the PTFPP identified the "seriously degraded catchment" of Tamlang as a landscape in need of comprehensive rehabilitation (PTFPP 2000, 3). Achieving the goals of the Catchment Management Plan in Tamlang (including Kamantian and Lap Lap) required that "further expansion of kaingin [swidden] areas . . . be stopped and [that] people should instead use existing grasslands" in traditional use areas (PTFPP 2002a, 6). Like Inogbong, most catchment projects pushed for forest rehabilitation through enrichment planting in swidden fallow areas and the afforestation of "denuded areas" (12). This involved an intensive awareness campaign on forest conservation through "signage and information-education-communication (IEC)" (12). Panglima and Barangay staff working with the PTFPP encouraged upland residents to undergo "non-agricultural based skills enhancement training for additional incomes to reduce their dependence on the forest for survival" (16). As chapter 6 shows, these organizational and leadership initiatives helped cement various social contracts between panglima, the community, and the PTFPP's biopolitical aims in the upland management zones.

Kamantian served as a strategic staging point for PTFPP initiatives. In the restricted-use zone, PTFPP staff first targeted Pala'wan resin harvests in the uplands, drawing on the ecological labor of panglima and villagers to reduce almaciga (bagtik) harvests to three kilograms per tree at two-to-four-week intervals. Next, staff aimed to reduce the rattan harvest, worrying that high harvesting levels would eventually deplete stocks. At a landscape scale, five hundred hectares of brushland needed to be restored by leaving lands fallow for ten years, another fifty hectares of *cogon* grassland needed to fallow for three to five years and be planted with "high value fruit trees and cash crops" to enrich the landscape, and, finally, a degraded traditional-use zone needed a sizable three-hundred-hectare tree farm for "improved kaingin" practices (PTFPP 2000, 13). The PTFPP staff justified this catchment-scale intervention by noting that "it is not possible to prohibit kaingin immediately, but it is possible to aim for its replacement in the medium-term by more sustainable land uses. Until that happens, the food needs of the families in the middle of the catchment will continue to be met by kaingin and the challenge is to educate and train the farmers in the guidelines already introduced by the PTFPP to improve kaingin practices" (PTFPP 2000, 17). Within this sociospatial fix, the Pala'wan of Kamantian and its surrounds had increasingly less space within which to harvest NTFPs and cultivate swiddens, as they had once done with greater freedom.

The PTFPP's reform agenda was comprehensive and totalizing; initiatives designed to protect upland forests and landscapes were connected to Indigenous health and education. Drawing on racialized, neo-Malthusian assumptions, the PTFPP argued that as Indigenous families overcame poor education, health, and morbidity, they would have fewer children and cultivate less swidden. The organization attempted to realize this objective by paying Pala'wan to construct permanent housing and other shelters in the lowlands and upland areas where PTFPP staff provided malaria bed nets, water pumps, and formal schooling (teaching subjects like spelling, hygiene, and basic ecology). The community catchment projects thus enabled deeper investments in the program's core logic of sustainability and modern living (PTFPP 2022e), which would translate into sedentary, forest-loving, and God-fearing farmers.

Enter Tingkep: Master Weavers and Hygiene

By the early 2000s, PTFPP ideology and infrastructure were already embedded in Kamantian proper, with fifty-two "forestry microprojects" being implemented in and around Tamlang's new ECAN-defined watershed area (PTFPP 2002c). Pala'wan in Kamantian were given additional training in how to weave better-quality handicrafts and how to better negotiate with the middlepersons who bought them in the uplands. The PTFPP soon hired Pala'wan staff from lowland sitios, such as Amas, to work with the residents of Kamantian to enhance their marketing opportunities, terms of trade, and profit margins for specific handicrafts (PTFPP 2002d). One basket in particular—the charismatic tingkep—was pivotal in leveraging sustainable livelihood schemes among the Pala'wan.

PTFPP staff went to considerable lengths to intensify the production and exchange of tingkep baskets. They smoothed the supply chains between upland villages and lowland towns and cities, compelling staff to establish better networks between tingkep weavers and "retail outlets in Puerto Princesa City"; they also noted the "limitations to the movement of inputs into their areas and the transportation of produce to market" (PTFPP 2000, 9). Newly constructed facilities in lowland and upland areas were staffed by market specialists who helped the weavers to "upskill" and craft smaller baskets with unique weaves. Spacious storage sheds were constructed—and continue to be used today—from local forest materials at the lowland trailhead leading to Kamantian and in Kamantian proper to store surplus baskets. These baskets were secured in the storage sheds before Pala'wan sellers transported them to Puerto Princesa City for final sales (see chap. 6). NTFP handicraft production soon expanded to include tiger grass brooms (*walis tambo*) and variously sized baskets (e.g., tiny tingkep) for more discerning tourists.

By this time, PTFPP staff had already noted an impending shortage of local NTFPs for weaving baskets and carving face masks out of local hardwoods: "There is no legal source of timber for handicrafts or building materials. Handicraft manufacture is being held back because of this. Tree farming represents the best way of overcoming this shortage and introducing a suitable land use for the communities in the traditional use zone of the middle catchment [that would not further exacerbate deforestation]" (PTFPP 2000, 16). The decision to prioritize tree farms over swidden fallows as the main source of materials for crafting pointed toward an integrated strategy to reform upland livelihoods well into the future. Comprehensive afforestation initiatives would stabilize swidden agriculture, supply timber for homes fixed to lots, and provide fiber for intensified "value-added" handicraft production. The PTFPP thus provided the platform for other NGOs, such as CI, to subsequently invest heavily in basket weaving as an alternative livelihood and income supplement—an apparent quick fix for poverty reduction and forest conservation. Ironically, most handicrafts were made from NTFPs like vines, rattan, and bamboo from old swidden fallows, rather than timber from tree farms.

NTFPs and Health Reform?

Part of the reason for the slow pace of development within the IP [Indigenous peoples] communities is the restricted access to schools. Literacy is essential if the IPs are to obtain access to better jobs. Schools also represent a means to extend the message of the PTFPP.

—Tamlang Catchment Plan (2000)

Externally designed alternative livelihood programs in Kamantian soon incorporated deeper health and educational reforms. Alongside handicraft production, contour cropping, and enriched fallows, PTFPP project officers requested that tribal-friendly lowland schoolteachers and health workers relocate to the newly built facilities to teach literacy. Advancing upland literacy, health, and agriculture were upheld as modern biopolitical solutions to overcoming the "primitive" and "backward" ways of the Pala'wan. Improving the poor health and hygiene of Pala'wan would apparently optimize "the productivity of households in the middle catchment villages" (PTFPP 2000, 18), where "part of the problem can be addressed through better education and provision of clean water to prevent illness" (18). The PTFPP organized regular visits from the rural health unit of the lowland Local Government Unit (LGU), complete with "organised medical and dental check-ups, health education classes and monitoring of nutritional status" (20). All Kamantian residents received leaflets and were shown videos (on portable players) about appropriate health and hygiene practices.

Although the PTFPP closed in 2002, the social and physical imprint of its reforms lingers to this day. Intensifying and optimizing the social and biological existence of the Pala'wan required that they: a) consider living permanently on less land and doing more with it; b) use fewer forest resources in fixed areas owing to restricted access to "core zones"; c) plant trees to stabilize swiddens in other zones; d) rely on "efficient" biomedical remedies and associated Christian beliefs, education, and calendars rather than customary healing practices and rituals; e) organize themselves to perform the ecological labor that would typically be the remit of government; and f) increasingly reorient familial and village material culture (i.e., tingkep) into intensifying commodity supply chains that feed the burgeoning tourism market in Puerto Princesa City. Yet the PTFPP was not the only biopolitical creature lurking in the shadows of upland spaces; by now, well-funded NGOs like Conservation International had already fixed their biopolitical gaze upon the Pala'wan highlanders. The parastatal, PTFPP, had simply smoothened the path for them.

Conservation Agreements and Direct Payments

The regional office of the global environmental NGO CI was established in the Philippines in 1995. Its in-country office was set up in Manila, but, like other prominent NGOs, it soon opened a satellite office in Puerto Princesa City. CI headquarters in Arlington County, Virginia, ultimately directed the broader vision and management principles of both offices. Not long after starting work on Palawan, staff in both locations were instructed to prioritize the management and protection of the endangered biodiversity hot spot, the Mount Mantalingahan Range, toward "Our Shared Goal: No More Forest Loss" (CI 2021).[7] In 2009, with help from foreign donors, CI Philippines and the Philippine government drafted a management framework to establish the expansive 120,000-hectare Mount Mantalingahan Protected Landscape (MMPL), the largest terrestrial protected area on Palawan.[8] As part of CI's market turn (Milne 2022), the NGO took the first step toward capturing the net value of the MMPL's forest in terms of "natural capital," with an imputed market value of "US$5.5 billion worth of ecosystem services." The ecosystem services were intended to benefit the upland and lowland peoples of the range (CI 2021).[9]

With national and provincial government financing, CI Philippines and the newly formed South Palawan Planning Council (SPPC) were tasked with implementing the new MMPL Management Plan (2010) and upholding the lofty objective of "zero net loss of forest and ecosystem services in the protected area" (CI 2021).[10] The SPPC technical committee and protected-area management board consisted of provincial and local politicians and Indigenous representatives from the five municipalities within the protected area's boundaries. Under CI's watchful eye, it aimed to conserve flora and fauna by mobilizing Pala'wan communities, offering alternative livelihoods, and facilitating educational programs. Like the PTFPP, these management initiatives sought to "reduce resource extraction to a sustainable level by supporting efforts to lessen communities' dependence on activities that degrade the environment" (MMPL 2010, 29). Achieving these management objectives meant drawing on the labor of Pala'wan forest guards (*bantay gubat*) and wildlife enforcement officers strategically located in lowland and upland villages along the protected area's boundaries.

Because of CI Palawan's direct role in establishing the MMPL, most of its upland interventions in the protected area adopted the much-lauded Community Conservation Agreement (CCA) approach. CCAs worked as binding social contracts with uplanders, "enabl[ing] and encourag[ing] local communities to protect their natural resources through conservation agreements, which incentivize community protection activities" (Conservation International 2021, 1). The NGO seldom questioned the transformative potential of CCAs, noting that, "to date, hundreds of families have taken part—and learned the value of conservation along the way—transforming communities from resource users into responsible and sustainable resource managers" (1).

CCAs worked as direct payment approaches designed to overcome the perceived failures of indirect community conservation initiatives—for example, integrated conservation and development projects (Brandon and Wells 1992)—that apparently lacked the incentives needed to significantly effect behavioral change (Milne 2009). This marked CI's shift away from integrative approaches and toward direct financial payments to local and Indigenous resource users ("village residents") for "increased effectiveness and efficiency in conservation investments" beyond the life of the agreements (145). CI's new market mantra entailed conservation payments and performance-based compensation for "measurable indicators" in the nonstate spaces of Palawan's interior (145).

The CCA model directly linked conservation to the apparent benefits of ecosystem services. The direct payments ostensibly incentivized Indigenous compliance and compensated for the livelihood opportunities lost after discontinuing certain resource uses, overcoming any assumed conflicts between "people and nature" (Milne 2009; Niesten et al. 2010). As the former senior director of the Conservation Stewards Program wrote, CCAs delineated rights, responsibilities, and associated benefits to "compliant" Indigenous resource uses and sanctioned noncompliance through the "graduated and temporary reduction in the benefits to allow for improved compliance and restoration of the full benefit package" (Niesten et al. 2010, 6). Thus, CCA direct payment schemes ultimately reflected market-oriented mechanisms, and compliance would somehow allow Indigenous uplanders to realize the full market potential of intact forest "ecosystem services." In more pragmatic terms, CCAs supposedly incentivized greater accountability to ensure adherence to measurable forest conservation and legitimized the social contract between upland communities and the NGO in remote forest settings. The CCA had become well-honed disciplinary tool.

A former director of CI Palawan, whom I interviewed in 2001, described how, from 1995 onward, the organization focused on "making parks work for the support of biodiversity conservation hot spots" through appropriate financing (CI interview 2001). During the interview, they emphasized the significant labor and financial resources that CI had invested in many of its projects, reporting that many Indigenous uplanders showed initial enthusiasm to participate, but were less inclined once the project's perceived benefits—including cobenefits and direct payments—dried up. As the former director cynically lamented, "CI comes with an institutional handicap from being perceived as having the money. So, people who would not know about the context of the project, they think that CI has lots of money to give around."[11]

In July 2009, CI Palawan implemented a CCA in the upland villages of Marenshewan, Bataraza, around sixty kilometers south of Kamantian's Tamlang watershed. Working with a Pala'wan panglima, "tribal council," and families, CI selected Marenshewan on the premise that "kaingin [swidden] was deviating from the traditional practices . . . [and so] is a major threat to the remaining intact forest in the . . . strict protection zones" (CI factsheet, n.d., 1). Countering this meant transforming farmers into paid "community rangers [who] in coordination with the local peoples would monitor the area to ensure old-growth forest and threatened species would not be cleared for swidden and harvested for consumption or sale" (1). CI incentivized the agreement's ecological labor by facilitating tingkep weaving and trade in the community, low payments to plant tree crops in swidden fallows (1), and other "cobenefits." Initiatives to commodify the tingkep and other handicrafts centered on incentivizing weavers to produce more baskets to feed newly established supply chains that would get the baskets to market more quickly. Expanding production and supply chains would supposedly increase basket sales and income and presumably improve uplander compliance with conservation strictures within the protected area.

More recently, CI expanded its interventions into the MMPL's uplands while assisting the DENR, PCSD, and US Agency for International Development (USAID) in the implementation of the Protect Wildlife Project across Palawan. The project was financed by payment for ecosystem services (PES), USAID, and the government (via tax levies from extractive industry, and tourism). It established so-called forest land use plans (FLUPs) comprising upland management plans, interventions, and zoning of protection, conservation, and production areas, which were supposed to align with and harmonize ECAN zoning and the SEP law. The $22 million US project used CI's initial natural capital valuation of the MMPL (Dressler et al. 2018) to ascertain the social and economic value of key ecosystem services through the FLUP and PES valuation architecture. This would supposedly finance lowland and upland Pala'wan communities pursuing sedentary "conservation agriculture" and "agroforestry" and generate income to help offset the need to cultivate swidden and hunt wildlife in the mountains (DAI et al. 2021).[12]

In time, Pala'wan residents were drawn into various missions—environmental and religious—and negotiated multiple, intersecting biopolitical reforms. As much as the tingkep drew families together to nourish the basket's vitality in myth and ritual, it was now being produced for distant markets owned by others. Such customary objects (and associated livelihoods) were soon reproduced as exotic

touristic commodities and harnessed by NGOs to incentivize forest conservation and induce sedentarization. As the Pala'wan dealt with similar PTFPP and
CI reforms, those in Kamantian were also contending with the AFM's attempts
to undo their beliefs and behaviors from the inside out. Evangelicals undertook
more insidious biopolitical reforms to suppress customary diets and objects as
"pagan" and "primitive."

The Deepening: Evangelical Biopolitics in the Uplands

The SDA had been present on Palawan since at least the 1950s, but the self-
financing, somewhat independent AFM camp in Kamantian was only established
in the early 1990s (approximately 1995), during the rise of environmental and
Indigenous rights NGOs.[13] An American pastor and his wife took over the mission in Kamantian between 1994 and 1995 to pursue a much deeper and comprehensive biopolitical strategy of attaining religious "purity and progress" among
Indigenous peoples across southern Palawan's interior. After graduating from
religious institutions in the United States and honing his seminary skills preaching at SDA conferences, the pastor and his wife came to learn about the AFM and
then joined as "church planters" (AFM 2021a). For the pastor, planting faith in
God among the Pala'wan was the surest way to fulfill His divine will (GMITV,
ND). Managed by the pastor and his wife for the last two decades, the mission
has regularly hosted young, ambitious American volunteers known as "Frontier
Adventists" who replenish the Kamantian camp to fulfill "God's mandate" to
"Reach the Unreached for Jesus" (Guerrero, 2010).[14] The official AFM website's
"Why We Go" section details the mission's core mandate:

> Our sole purpose is to hasten the coming of Jesus by giving unreached
> people the opportunity to know Him and become His disciples. We
> are the only Seventh-day Adventist ministry focused solely on the
> unreached. God's instructions are clear: His people must take the ever-
> lasting Gospel to every nation, tribe, language and people (Rev. 14:6).
> Then the end will come.
>
> • There are nearly 7,000 unreached people groups representing
>   more than 2 billion people.
> • Out of every 1,000 Gospel workers worldwide, only 14 serve
>   among the unreached.
> • Out of every $100 given for Christian ministry worldwide, less
>   than one penny goes to help reach the unreached.

Is this status quo acceptable? Absolutely not! Every day, thousands of people die without any knowledge of Christ or the salvation. He gave His life to purchase for them. The unreached are waiting for someone to tell them.

Who are the unreached? They are people groups, often isolated behind barriers of geography, language and prejudice, who have no access to Gospel truth because there is no viable Christian presence among them. (AFM 2021b)

Young missionaries with specialized skills (e.g., carpentry and nursing) useful in maintaining the camp infrastructure in Kamantian (e.g., health clinics, schools, churches) are recruited through the Southern Adventist University in Tennessee and other areas of the United States. They raise funds to cover the costs of the trip and receive a small monthly stipend for food costs (Southern Adventist University 2021). Under the pastor and his wife's direction, the young missionaries work patiently and persistently to expunge Indigenous faiths, beliefs, and practices, with the Bible and sermons delivered in the Pala'wan language. The AFM considers "Satan's presence" to be strongest among "interior Pala'wan" who all uphold substantive adat customs, including rituals involving tingkep (see chap. 7). The AFM's biopolitical approach is unrelenting; it works recursively and extensively in its bid to discipline "unreached [Pala'wan] people" further and further into the uplands:

AFM works among unreached people groups that have no Adventist presence. Once a solid foundation is established, our missionaries begin discipling and training new believers to carry the work forward.

This evangelism model not only empowers local believers, it enables the Seventh-day Adventist Church to grow and expand in areas previously unentered. Many of the areas where AFM works are highly resistant to the Adventist Church for social, political, or ethnic reasons. Our unique cross-cultural approach to evangelizing unreached people groups breaks down these barriers and allows for exponential growth. (AFM 2021b)

In Kamantian, the AFM missionizing I observed during field visits was less cross-cultural than deeply pastoral and biopolitical. Decades ago, Macdonald (1992a, 130) described the New Tribes Mission in southern Palawan as having an intermittent influence; most Pala'wan gravitated to the mission for "goods and services" and returned to "old customs and ritual practices" after lead missionaries had departed the area. By contrast, the AFM presence and impact in Kamantian appeared to be both enduring and deeply consequential for those

Pala'wan in the mission's evangelical fold. The young Pala'wan disciples I met spoke devoutly of the mission and its agenda among their people.[15] Mission life appeared to be a serious biopolitical affair; the pastor, his wife, and other younger Frontier Adventists not only studied Filipino ways, but aspired to understand Pala'wan language, social relations, and customs. The AFM used this knowledge to delegitimatize and overcome panglima, beljan, and other Pala'wan leaders and forge enduring Indigenous disciples—a practice not too dissimilar from CI and PTFPP. Pala'wan Adventists also regularly visited their extended families to build trust and enrol them in church services. AFM proselytization was done from both the outside in and the inside out.

Such incremental disciplining involved an array of well-structured biopolitical practices: linguistic and translation work, profound conversions of individual Pala'wan trainees and other villagers, and, ultimately, the shepherding of local leaders to maintain the mission. This approach—the AFM's "church-planting model"—was both incremental and enduring in overcoming paganism and animism. Long-term, village-based engagement was critical to these reforms. The AFM's website describes the evolution of its church-planting methodology:

> In nearly 30 years of ministry to the unreached, AFM has been constantly developing and improving its model of effective cross-cultural church planting. Recently, we developed a new refinement of our model to guide our missionaries through key phases of their ministry.
>
> The new AFM Church Planting Model is not intended to be linear. Several aspects contained in the main circle of ministry can be taking place at the same time. (AFM 2021c)

Following this approach, the AFM "expects [evangelical] multiplication" in the uplands: "When we expect a tornado, we prepare for it. When we expect that a church committed to God will want to multiply, then we should work with God to prepare for that as well. This all starts with fervent and constant prayer each step of the way. And as the disciple-making cycle repeats continually, the project grows as directed by the Holy Spirit. Believers are organized into multiplying churches as they reproduce the cycle with new members" (2021c).

By studying Pala'wan social life, the Adventist Frontier missionaries aimed to identify and meet community needs and concerns to entice Pala'wan into the seemingly beneficial fold of the faith (e.g., medical care, livelihood "support," and labor opportunities). Through church assemblies and prayer, they then introduce more and more Pala'wan to the faith and identify those with leadership traits. Potential Pala'wan leaders are nurtured through "leadership development" programs and tasked with spreading the faith among close-knit kin networks—a strategy that mirrors the PTFPP's capacity-building strategies. Chosen "future leaders" would also endear themselves to the young Adventist missionaries, spending time in their company and expressing their devotion. The most prominent of the Pala'wan AFM leaders would be enlisted as "maturing disciples" to continue the upland mission. Finally, follow-up initiatives would ultimately support the AFM mission with financial assistance and communication with senior church leaders who visit the camp.[16] As I show later, this process perpetuated a cycle of recruitment and socialization for Pala'wan "believers" and the proliferation of Adventist churches in the southern uplands. The AFM would build upon the PTFPP infrastructure in Kamantian to reinforce adherence to faith and behavioral changes through contrasting (but often very similar) techniques, conditions, and strictures that aimed to constrain possibilities and optimize life among many, and resistance among few.

Like those of the PTFPP and CI, the AFM's biopolitical strategies drew on the racial logics of cultural difference to distinguish between their target population and others (Bierman and Mansfield 2014, 258). Working through racial significations of what constitutes "normality" and "abnormality" in faith, social practices, and diet (Bierman and Mansfield 2014, 258), the AFM drew on techniques and practices to punish, incentivize, and deeply reform upland residents in their homes and community. Although the AFM aimed to fully convert and subjugate the Pala'wan to scripture, the pastor and his disciples also understood that this reform process would be (at least today) partial, incomplete, and contested. In this sense, they knew well that their work would require patience and persistence to ensure that the mission's architecture was firmly established and resistant to dismantling and defection. Emerging disciples knew that any deviation from morally sanctioned behavior would be socially reprimanded and that further "backsliding" would lead them to the "gates of Hell." AFM biopolitics thus worked to align the Pala'wan with a deeper pastoral care by offering them incentives and opportunities relative to the needs and constraints of upland life. This included the treatment of illnesses in clinics or offering them paid labor to support the AFM. As the pastor argued, this is the thin end of a wedge needed to make the Pala'wan see "the light of God."

This chapter has provided historical context for how the logics, practices, and motives of various nonstate actors have intersected as they fixed their biopolitical gaze on upland peoples and forest spaces in southern Palawan. Since the early 1990s, a range of international and local nonstate actors have applied different interventions with the aim of disciplining and reforming the tribal Other; the desires and practices of such reforms have long been informed by colonial histories, land laws, and racialized categories of difference that still flourish today. For many nonstate actors, pagan lifeways and degrading livelihoods remain powerful beacons of reform—signals to implement religious purity, ecological sustainability, and overarching modernity.

The practices of reform adopted by the PTFPP, CI, and SDA all converge around the desire to enhance and improve the apparent deficiencies that are considered endemic to upland peoples and landscapes in Palawan and elsewhere in the Philippines. Each actor's interventions have been anchored in colonial histories and contemporary upland realities that supposedly call for new market-oriented incentives and agreements. Getting the job done increasingly required minimum direct payments to ensure that Pala'wan ecological labor and forest uses aligned with newly forged conservation agreements, motives and futures. Broadly independent of the state, the PTFPP and CI used critical area zoning to regulate permissible land uses across ancestral lands, delineating core zones to enclose older forests for preservation and introducing programs to convert forest mosaics into enhanced, stabilized fallows in bounded traditional use areas. The imposition of such zoning and management upon Pala'wan fallow lands generated a sociospatial fix in the uplands, severing customary rights to land and forests upheld through rotational clearing, burning, and planting crops across diverse landscapes. In their place, entrepreneurial forms of agroforestry and tingkep production for touristic consumption in lowland markets were encouraged as "sustainable" livelihood alternatives—interventions that planted biopolitics that supplanted Indigenous rights to land and livelihood.

The AFM built on this biopolitical matrix but significantly transcended environmental nonstate reforms with much deeper, comprehensive pastoral reforms. Over several decades, the AFM established itself deep in the Tamlang catchment to identify and convert Pala'wan "suffering from Satan's grip" in Kamantian and other remote sitios. AFM pastors worked as sanctioned "shepherds" who disciplined emerging "prayer warriors" that would regulate the village population as their flock (Foucault 2003, 249). They used their medical clinic to heal, incentivize, and build trust, while deploying religious doctrine and strictures to prevent Pala'wan backsliding. Each nonstate actor has subsumed branches of government in the uplands, unleashing biopower that, in Foucault's words, is seemingly "a beneficial power . . . its only *raison d'être* is doing good . . . to ensure sheep do not suffer and do not stray off course" (2007, 127). In many ways, both environmental and religious nonstate actors function as overzealous missions that invoke a pastoral duty of care to ensure Indigenous uplanders overcome the assumed suffering and vulnerability of forest-based livelihoods and ways of life. These attempted reforms are implemented whether the Pala'wan request them or not.

As shown, nonstate agents work off one another's interventions and infrastructure; they almost always use similar logics, motives, and disciplining practices. Moreover, their biopolitics involve diverse multiform tactics that similarly aim to reform social reproduction across the public and private spheres of the individual, family, and village (Li 2007). In upland areas, nonstate biopower thus works through the *intimate and personal* spaces of Pala'wan lives and their customary objects, directing them to live better, to live life optimally, and to live in line with modern norms and presuppositions.

As I show next, these social reforms are often mediated through socially and materially important objects such as the tingkep. Environmental NGOs created a fledgling basket weaving industry to generate additional income and offset livelihood opportunities lost because of prohibitions against swidden cultivation and the transition to sedentary livelihoods. In contrast, the AFM prohibited the use of customary objects such as the tingkep for rituals and ceremonial practices (e.g., carrying powerful mutja to imbue a hunter with strength and courage). However, many uplanders have distanced themselves from the barrage of interventions aiming to reform them; many have become despondent; others continue to resist in subtle ways.

The next two chapters pivot to ethnographically examine nonstate biopolitics in practice. They trace the converging influences of these interventions and describe how multiple project techniques and practices converge in the uplands, with a range of intended and unintended consequences for the Pala'wan. I foreground how the tingkep mediates these changing nonstate interventions and practices across upland and lowland settings.

FOR THE SAKE OF FORESTS

> Sometimes I think that it was much better before CI came. Before when we had kaingin it allowed us to survive; if we don't have kaingin I feel that we've lost everything. We had big hopes, but now there's nothing. That's why we only get money if we sell tingkep, plant fruit trees from the nursery, and if we can work for them, we can get money; but if there's no environmental work there's no money, and we only eat cassava.
>
> —Panglima Bilog, CI broker

Panglima Bilog's statement reflects the contrasting motives and interventions emerging from nonstate actors taking on de facto state roles and responsibilities in their attempt to govern Pala'wan life and livelihood. Using similar interventions, the parastatal PTFPP helped establish the governance architecture for Conservation International (CI) to pursue programs that mixed sanctions and incentives to actively "render problems amenable to intervention" (Li 2016, 79). Their projects, guided by totalizing agendas, worked strategically through the sacred and profane of everyday life in the uplands of southern Palawan. This chapter examines these nonstate *practices of reform* over Indigenous peoples' forest-based existence, from the everyday of making a living to the sanctity of ritual practices. It explores how PTFPP and CI staff and programs exercised biopower as routine practice, how these practices were entangled with livelihood production and exchange in the uplands, and how the Pala'wan responded to reform agendas across rural and urban Palawan.

I focus on how the social labor and contracts of such projects worked through family and village social relations and material objects—whether the customary tingkep or the pragmatic machete—to sedentarize livelihoods and emancipate custom, irrespective of the consequences (Keane 2007). The nonstate labor involved in designing and enrolling the Pala'wan into projects increasingly centered on market-oriented social contracts that encouraged communities to participate in and accept "conservation rule" in exchange for receiving project benefits and opportunities (O'Leary Simpson and Pellegrini 2021, 1–2). Such conservation social contracts reflected "a form of agreement . . . , formal or informal, between two or several parties, which involve[d] the regulation of the extraction, consumption and trade of natural resources" (Titeca et al. 2020, 3). They exist between "state and citizen [but increasingly] connect multiple [nonstate] actors that operate across different hierarchical levels, and shift in time with broader social, political and economic relations" (Simpson and Pellegrini 2021, 3). Nonstate actors design—whether as blueprints or in tailor-made form—social contracts to socialize and realign rural communities to "manufacture new aspirations and behaviours, to convince rather than coerce" in remote forest settings (2). The imperatives of social contracts are thus deeply biopolitical.[1] Their clauses expect resource-reliant highlanders to forgo clearing and burning forests for swidden or hunting threatened species in exchange for "stable" and "sustainable" livelihoods, often with few tangible benefits.

In the southern uplands, nonstate actors are drawn to specific social registers—degrading Pala'wan swiddens and hunting, customary objects, and ritual practices—and interpretations that inform contract designs and embedded desires for enduring social reforms. As Ferguson (2006, 178) notes, "The extent to which societies differed from the modern ideal neatly indexed their 'level of development' toward that ideal." Nonstate actors indexed and repurposed contrasting cultural values and aesthetics of certain social and material relations through racialized categories. The tingkep and its uses emerged as one such register, repeatedly repurposed to reform human-nonhuman relations. However, the tingkep's relational character and usage, as customary object and exotic souvenir, also mediated disciplinary influences upon Pala'wan livelihoods and behaviors to align (or not) with socioecological ideals across rural and urban settings. Indigenous objects and relations mediated efforts of reform, even when valorized or demonized by nonstate actors. In the same highland spaces, then, the converging reforms that marked social and spatial differences also had difficulties flattening and overcoming them.

This chapter is divided into three parts. The first part describes how the PTFPP incentivized Pala'wan to make tingkep as an authentic "native" souvenir to draw income from tourism sales and possibly offset the need to slash and burn forests. In the second part of the chapter, I introduce how CI replicated the PTFPP's logic and style of intervention through its market-oriented trademark, Community Conservation Agreements (CCAs) in Marenshewan in the uplands of Bataraza. While CI used many cobenefits (e.g., direct payments for tree planting), tingkep production emerged as an important disciplinary strand woven into localized practices of reform among the Pala'wan. I then show how the PTFPP and CI regularized access and use restrictions through the related livelihood incentives and strictures of expanding conservation zones. Finally, in the third part, I describe how, partly because of these project interventions, more uplanders began selling more baskets and other crafts in the capital city, Puerto Princesa, feeding traders, sellers, and buyers. A greater number of Pala'wan basket traders soon made the long journey from their upland homes to the city, where they tried to sell their crafts to different buyers. I describe their experiences with souvenir shop owners, who denigrated Pala'wan material culture even as they bought their finely crafted goods. Ultimately, the touristic desires for traditional upland objects and souvenir shops' hoarding and (mis)representations reinforced the Othering of the tingkep and, ultimately, the Pala'wan themselves, echoing colonial racializations.

Nonstate Legacies: The Markets and Ecological Reforms

By the late 1990s, the PTFPP's afterlife and CI's legacy had left discernable imprints in both Kamantian and Marenshewan. Valorized leadership roles, environmental education, sanctions and zones, economic incentives, and infrastructure were in place to subsume forests, livelihood needs, and everyday politics. This nongovernmental assemblage—the work of scientific expertise, technical knowledge, and striving to identify problems, diagnose them, and propose solutions—had replicated itself and intersected with life in the Palawan highlands (Li 2016, 80).

In Kamantian, the Pala'wan no longer used the tingkep solely for livelihood and in customary practices but increasingly as a material object, a partial commodity, to generate income to offset the costs of declining crop yields or not clearing older forests for better swidden yields. By this time, at least ten families were busy weaving baskets as souvenirs to meet the rising demand for "authentic native crafts" from domestic and international tourists. While baskets were woven almost constantly, most were made during the rainy season or in the warmer afternoons when less time was spent in swidden fields and forests. An expert weaver in Kamantian could take three to five days to weave smaller tingkep and seven to ten days to weave a larger tingkep or a tabig. While the larger, sturdier baskets were still made to store household items such as rice, root crops, clothes, or ritual objects, the smaller, often thinner, tingkep were increasingly becoming commodified ornamental souvenirs. The very best weavers resided in Kamantian and Marenshewan, making the upland sitios the main source locations for the baskets and other customary items.

In the early 1990s, only a few itinerant traders from Puerto Princesa City (PPC) and Manila (e.g., Eliza Salazar and Irma Abueg), who worked independently of nonstate actors, traveled to purchase tingkep from Pala'wan weavers high in the Mount Mantalingahan Range. At that time, traders sold the baskets to just two or three souvenir shops in PPC. Only a few thousand tourists visited Palawan annually, and very few traveled south. In the early 2000s, however, the nature and scale of tingkep production changed dramatically. As tourist numbers and demand increased, Pala'wan weavers produced more baskets for the PTFPP and CI, who extended the basket's supply chains to a growing number of souvenir shops in the city. Many more Pala'wan men and women began acting as middlepersons, collecting, buying, and selling the baskets from households across the uplands and hawking them at urban markets or to residents along the hot, dusty roads of the city. This multistage journey involved collecting baskets and other handicrafts from upland villagers, traveling to the lowlands and road-sides, loading goods into "jeepneys" (local jeep-trucks) for the three-hour drive north, and spending several days in the city center (and periphery) selling the crafts. This journey was arduous and resource depleting. Back in the uplands, Pala'wan households harvested more nontimber forest products (NTFPs) to intensify tingkep crafting and meet the growing demand, ironically contributing to resource declines.

Tingkep Intensification, Markets, and Ecological Reforms

Souvenir tingkep production began in earnest in 1996 in the Kamantian area. Pala'wan broker (and buyer) Sofrano Aguilar learned how to optimize the weaving, buying, and selling of tingkep from earlier traders and PTFPP livelihood projects. Initially, Sofrano and his mother—an expert weaver—sold their baskets to the lone migrant trader Eliza Salazar, who had established an expansive tingkep-sourcing network in Kamantian, Marenshewan, and villages near Quezon and Rizal along the eastern and western ridges of the range. Eliza drew on entrepreneurial skills developed in the Visayas to create an outlet that sold and stored tingkep and other customary items along the national highway to PPC. She initially named the business the Kamantian Handicraft Shop but later rebranded it as a "cooperative." Two decades later, the shop relocated to bustling Rizal Avenue in the city, the area most frequented by domestic and international tourists seeking souvenirs.

According to Sofrano, it was about 1998 when the terms of trade between Eliza, his family, and Pala'wan weavers began to sour. While Eliza had a registered license to buy and sell the baskets in Puerto Princesa and Manila, her payments were too low, leading many weavers to incur debts from cash advances provided to complete orders. Sofrano decided to take matters into his own hands. Utilizing the lessons learned from Eliza, particularly his knowledge of existing producers and trading networks, Sofrano began trading baskets and other handicrafts on behalf of his people in Kamantian.

In time, fellow Pala'wan and then–PTFPP staff member Artisa Mandaw from the Brooke's Point satellite office was tasked with launching the Tamlang Catchment Management Plan in Kamantian. After completing a Livelihood Appraisal and Product Scanning (LAPS) initiative, Artisa realized that the tingkep had untapped potential as an "Indigenous heritage product" and that it could broker "buy-in" for the conservation program. According to Artisa, "Sofrano's tingkeps were just idle," so she helped him match "local skills and designs to the market." Sofrano also spoke of this training: "Yes, the [PTFPP] told me to ensure the weaving was sustained, especially the quality and buyers' interest in the product. They also taught me how to become better at buying and selling." In contrast to the intermittent sales by him and his mother, the PTFPP showed Sofrano and other buyers how to sell to recently established souvenir shops in PPC. He recalled that both Eliza and the PTFPP "helped me get to know those PPC tingkep buyers at the best souvenir shops. The PTFPP connected us up." With the PTFPP's assistance, Sofrano took over—in quasi-monopolistic fashion—the expansive tingkep territory from Kamantian to Marenshewan.

Artisa then introduced Sofrano to Pesti, the owner of Culture Shack, a well-known PPC souvenir shop, to conduct the "very first costing, sizing, and texture analysis of the tingkep ever." Pesti helped Sofrano formalize his business by setting up a bank account, securing a registered license, and expanding his trade from the forested uplands to Manila and the world. Robert Lane, the celebrated curator of Manila's Galeria de las Islas, soon also succumbed to tingkep fever, with the PTFPP asking him to facilitate workshops at Brooke's Point aimed at improving the quality of the basket, which Sofrano enthusiastically attended.

As a result of these interventions and his own initiative, Sofrano scaled up his own buy-and-sell operations from central to southern Palawan. He identified and ordered the highest-quality baskets (e.g., with clean, consistent weaves) from weavers in Kamantian and other villages across the uplands and sold them in bulk. Sofrano, his family, and others close to him had by now largely pivoted from laboring in swiddens to become dedicated entrepreneurs in basket production and exchange. Many had become expert supply chain enablers. From the PTFPP's perspective, this outcome was emblematic of an initially successful biopolitical intervention. Its job was not done, however, as most Pala'wan still wove, slashed, and burned simultaneously.

With his market expanding, Sofrano offered Pala'wan weavers cash advances for bulk purchase orders on behalf of the PTFPP (and Petsi) and other souvenir shops across the city. Following the PTFPP's advice, Sofrano began selling large quantities of tingkep and other crafts at festivals in the city. These festivals (e.g., Barakalan sa Tourismo and the large Baragatan festival) were sponsored by the city government's tourism office to celebrate "authentic" Indigenous crafts, food, and culture from Palawan. By this time, the baskets were also sold in high-end art stores as "tribal crafts" in Manila and as far away as Brussels, Belgium. It was in these spaces that branding the basket's "exotic," "customary," and "tribal" character reaffirmed the biopolitical basis for reforming uplander livelihoods toward sustainability.

PTFPP extension officers continued to promote tingkep and other handicraft production among uplanders as part of the Tamlang Catchment Management Plan. In 2002, a PTFPP (2002d, 30) report outlining "community micro-project implementation" stated that "marketing is for the sustainability of projects" and "a range of suggested basket sizes is indicated [but it is better] to pursue a small range of higher quality goods which have a high demand." The PTFPP and traders like Sofrano soon focused on enhancing weaves, producing varied patterns, and increasing the supply of smaller baskets.

As the PTFPP nudged the Pala'wan to produce more tingkep, the weavers were given saplings to plant in swidden fallows. The logic was that Pala'wan who received a regular income from intensified tingkep weaving and planted valuable saplings in recently harvested swiddens would be less inclined to clear and burn fallows in the catchment's core zone. PTFPP's complex infrastructure in Kamantian included a bunkhouse for visiting officials, a clinic for "traditional healing," and a schoolhouse for "alternative learning," but it was the coupling of overleveraged tingkep supply chains and tree planting initiatives through Pala'wan labor that was so central to reforming their livelihoods, education, and health practices. Built from Kamantian's forests, parts of this social and material infrastructure would eventually be handed over to the Adventist Frontier Missions (AFM).

Souvenir Supply Chains and Othering

As tourism flourished, a greater number of souvenir shops opened in PPC and sold the increasingly coveted tingkep along with a range of other crafts from Indigenous highlanders. Both NGOs and souvenir shops represented the basket as authentically "native"/*nativo* or "tribal"/*tribu* to influence demand for the exotic customary object in Palawan and across the Philippines.

Working under informal contracts drawn up between souvenir shops and traders, Sofrano, other traders, and local fixers commissioned and collected significant numbers of different-sized tingkep from Kamantian and surrounds. As connections with souvenir shop owners expanded in PPC, traders visited Kamantian and nearby areas every few months to collect their orders, often securing thirty baskets or more at a time from one village. The stock of baskets was stored in a thatch hut (*kubo*) storehouse near the trailhead to Kamantian (and elsewhere) and was replenished regularly. When sales slowed, the storehouses would overflow with baskets and other crafts waiting for collection and future sale.

Sofrano often worked with known expert weavers. In 2013, he tried to pay all Pala'wan in Kamantian the same price of 70, 100, and 200 pesos (approximately US$1 = 45 pesos [₱]) for small, medium, and large tingkep, respectively. He then sold the same-sized baskets to souvenir shops for around ₱100, ₱150, and ₱300—a modest markup. Souvenir shops sold the baskets for double the

FIGURE 6.1. Weaving smaller tingkep for traders, 2013. Photo: Dressler.

purchasing price (e.g., small tingkep sold for ₱300–350). In time, more Pala'wan learned how to collect, buy, and sell tingkep and other handicrafts from fellow villagers across the southern ranges. Most self-styled Pala'wan traders who sold baskets in lowland towns and PPC were men with a degree of social standing or authority in upland villages, often panglima or respected elders.

Upland Pala'wan families received cash advances from Sofrano for medium (ten or so) and larger-sized (more than twenty) tingkep orders that helped cover livelihood costs, particularly sundries and rice at the end of the swidden season (when rice stores are lowest or depleted). Pala'wan households, typically families with several hungry young children, often spent their advances quickly on lowland rice, only to request more cash while promising to repay initial installments by weaving more baskets. However, few achieved this goal; as families produced more baskets on advances, they often slipped further into debt, with some owing as much as ₱1,000. As Pala'wan weaver Benecio Buat noted, "Before I had ₱1,000 in debt and to repay it sometimes, I needed to make twenty smaller pieces of tingkep. If we do not have another job, it can take us about six months to make that many, but if we have to prepare our swiddens [uma], it will take about eight months."

Livelihood Pressures and Outcomes

When Pala'wan families struggled to repay their debt, the traders' debt also rose, eating into their savings and capital (particularly when shops failed to pay).[2] Ultimately, the traders intimated that they could no longer offer advances to weaving families in Kamantian and elsewhere and so avoided buying baskets from those who failed to repay their debt.[3] Although the buy-sell relationship was never coercive, many families remained indebted and tied to Pala'wan brokers.[4]

Investing more time in basket production to cover debts incurred moderate opportunity costs for other livelihood activities. Pala'wan families generally engaged in a diverse mix of livelihood activities for subsistence (e.g., swidden, hunting, and honey harvesting) and income (e.g., rattan, almaciga resin, and wage labor); however, many women and children spent considerable time weaving baskets for bulk orders, often weaving for several hours at a time. A 2012 questionnaire (thirty householders, ten of whom wove consistently) that I administered revealed that, in the calendar year, some families had their children and cousins weaving up to 250 small tingkep baskets to meet large orders. Sometimes the year of weaving paid off, with certain households earning an estimated annual average income of ₱4,844 (US$100) producing smaller tingkep.[5] Overall, the average income from tingkep sales across all weaving households surveyed in 2012 was ₱3,518. While many considered crafting supplementary to other incomes, excluding debt and expenditure, this weaving income was higher than the average annual income for the more labor-intensive work associated with harvesting key NTFPs. In 2012, the weaving households only earned an average annual income of ₱2,680 from harvesting 272 kg of almaciga resin (*bagtik*), about ₱11–15 per kg for Grade B resin, and about ₱360 for 22 kg of honey (*deges*), which is about ₱16 per kg.

Harvesting bagtik is a physically arduous task. Collectors—able-bodied younger or middle-aged Pala'wan men—must walk upland to the interior stands of the resin-exuding tree (*Agathis philippinesis*). The small teams spend two or three days collecting resin at or near the base of the trees, depending on their location and elevation. They use machetes to make an incision at breast height, or higher with a makeshift ladder (Razal et al. 2013). Each tapper works on several trees and draws between eight and twelve kilograms of resin per tree. The resin is packed in a *kiba* (a pack woven of rattan), and the harvesters carry as much as forty to sixty kilograms of resin (often barefooted) down steep, wet, and slippery mountain trails to lowland *kapatas* (intermediaries) or other buyers.

Doing this work repeatedly places the harvesters' bodies under considerable strain and exacerbates underlying physical health conditions like tuberculosis. After deducting advanced loans of money and goods from the buyers, the sale generally provides little return on labor time invested. Many, therefore, bypass the kapatas and sell directly to buyers to receive a more competitive price. Nonstate actors are aware of this and the fact that the craft production is a seemingly less arduous, more lucrative "sustainable" livelihood activity, particularly for older Pala'wan. In most weaving families, however, it was mostly women and children who performed most of the—often arduous—weaving work, working portions of the day in their own homes, long into the cool evenings under the light of new (battery-powered) solar lights in Sofrano's home, or in the shelters constructed years earlier by the PTFPP (see figs. 6.2 and 6.3). In Kamantian, tingkep making had become a busy cottage industry spanning multiple villages and families. While the repeated physical strain of a fifty-kilo sack of resin differed from crafting, the sustained weaving strained the eyes and bodies of women, both young and old.

Weaving Overdrive?

Far from the PTFPP's "sustainable livelihood" ideal, the intensification of tingkep production across multiple villages soon depleted local stocks of the NTFPs needed for weaving (e.g., *busneg, buldung*, bamboo; *arurung*, rattan, *Calamus* spp.), compelling families to travel further to find suitable materials to keep up with demand. More entrepreneurial Pala'wan who wove tingkep intensively even hired other Pala'wan to collect NTFPs further upland. When arurung (rattan)

FIGURE 6.2.   Numerous finished baskets waiting to be carried lowland, 2018. Photo: Yayen.

supplies dwindled locally or were difficult to harvest during the rainy season, weavers placed regular orders. As elder Victoranio Itom noted, "If there is no arurung around here, we will make orders with other people who are collecting. But if we don't have any money, then we can't pay anyone to find arurung further in the mountains. Also, during the rainy season it's slippery and we old people can't climb the mountains anymore."

The growing problem of "scarcity" was also partly compounded by the difficulty of accessing NTFPs from nearby forbidden lihien forests—the nonhuman spaces that Linamen and Gila'en reside in and jealously guard. As Pala'wan

FIGURE 6.3.   Producing many smaller Tingkep for sale in the city, 2013. Photo: Dressler.

farmer Maria Siblang noted, "The *giba* [old growth] holds the Linamen, so we cannot use the area, for if she is disturbed, we will become seriously ill and possibly die." Accessing materials in lihien forests was often only possible after performing *ungsud* and receiving permission from the spirit world (Theriault 2017).

The PTFPP's Aftermath

My conversations with Pala'wan farmers and weavers in 2013 and 2018 made it clear that the PTFPP's "sustainable livelihood" interventions only further entangled the Pala'wan within the uneven commodity relations of lowland political economies. Sometimes this incorporation offered Pala'wan a predictable flow of cash income and food security, but it most often led to indebtedness and servitude. Such livelihood interventions responded to the PTFPP's biopolitical imperative to render Pala'wan into sustainable, sedentary subjects. Many Pala'wan elders worried that the associated "project packages" would take them away from swidden and upland life more generally. As one elderly Pala'wan man from Kamantian explained: "The PIADP (PTFPP) gave households land stewardship certificates to keep us in place.... They forbid us from clearing because it may cause a landslide, so right now there are already many trees planted there in the mountains by us. We are afraid of the PIADP, we cannot burn up there anymore."

The PTFPP's existing zoning systems, infrastructure, and disciplinary logics eventually offered other organizations political openings and legitimacy. CI (and AFM) would build upon and replicate what PTFPP had accomplished decades earlier by setting up their biopolitical camps south of Kamantian.

Social Contracts of Reform

If you don't cut down the old forest, we'll give you a chicken.

—Pala'wan farmer Mantap Bagrar

Further south in Marenshewan, CI Palawan was busy expanding its community-based conserved areas as part of its pledge to achieve "zero net loss of forest and ecosystem services" in the Mount Mantalingahan Protected Landscape (MMPL)—the same protected area it had helped established in 2009 (CI 2021). Noting the US$5.5 billion value of the MMPL's forest cover and "ecosystem services," the CI chapter set out to "provide paralegal training and enforcement mentoring sessions to over 45 [Pala'wan] community volunteers to represent a specific watershed. The volunteers [would be] deputized by their local chief executives and now conduct regular foot patrols in their designated areas" (2021). The NGO then aimed to enlist "300 [Pala'wan] families [to sign] conservation agreements to agree to protect their natural assets and learn the value of conservation" (2021), noting "plenty" of incentives. CI selected Marenshewan on the grounds that significant tracts of old growth forest in highland watersheds were supposedly being threatened there, noting in its CCA Progress Report 2009–10 (CI 2010, 5) that "continuous *kaingin* [swidden], deviating from the traditional practice, is a major threat to the remaining intact forest in the MMPL's strict protection zone." Marenshewan had become the NGO's test site for its trademark Community Conservation Agreement (CCA) initiative.

As a formal social contract, CI's CCA went beyond the PTFPP's institutional groundwork of supplementing and substituting Pala'wan livelihoods (Simpson and Pellegrini 2022). The CCA used direct cash payments and various livelihood cobenefits to incentivize sedentary practices like small-scale livestock rearing, while actively discouraging and regulating swidden and NTFP harvesting further upland. Achieving this involved enlisting Pala'wan participants in numerous biopolitical activities. They attended farm planning and improvement seminars, community volunteers were deputized to patrol and monitor the condition of forests and record any illegal activities, and many were paid to rehabilitate fifty hectares of "degraded forests" by planting native trees, at about four pesos per tree (CI 2021). CI hoped that the planting of higher-value tree species would discourage the Pala'wan from clearing and burning these "regenerative fields." If Pala'wan farmers breached the "mutually agreed" CCA rules by opening swiddens in mature forests, the various payments for their ecological labor would be withheld (Neimark et al. 2020).

Implementing the CCA among Pala'wan in Marenshewan and beyond required that CI appoint local agreement brokers at the organizational and community levels. At the organizational level, CI hired the congenial pastor Manong Pilmar to facilitate the CCA and its Environmental Literacy and Informal Education Program. As a pastor of the local Assemblies of God Church, an Evangelical Protestant mission, Pilmar was already familiar with the Pala'wan in Marenshewan that he was trying to recruit and baptize at his small church located just off the National Highway. He frequently traveled up to Marenshewan to help the vulnerable Pala'wan by spreading the word of God, lest they succumb to Satan's grip. With the trail to Marenshewan located behind his church and house, he let many curious Pala'wan gather by his residence to listen to his sermons or the radio or to watch television. Many Pala'wan knew Pilmar and trusted him. At the community level, CI hired the politically astute village panglima Franco Bilog as its main tribal broker and representative. Liaising with Pilmar and CI staff, Bilog's job was to enrol Pala'wan families living around Marenshewan and further upland into the CCA and ensure they aligned with its rules and regulations.

Unsurprisingly, it was Bilog, his three wives (two of whom were afflicted by leprosy), and his ten children who ultimately benefited most from the CCA. Fortunately for Bilog's family, CI's pastoral care—through Pilmar's religious environmentalism—involved carrying a regular supply of medicine to treat Pala'wan afflicted by malaria and, more urgently, the women's spreading leprosy, which had already consumed their noses, fingers, and toes. Bilog, who lived in a larger thatch house (with smaller "isolation huts" for his wives) near the trailhead to Marenshewan, used his strategic location and political authority to receive guests and broker deals, emerging as CI's emblematic tribal leader and forest steward (Murti and Buyck 2014).

CI first worked with the Marenshewan "tribal council"—a notional council created on the basis of imaginaries of indigeneity and its associated legalities (Indigenous People's Rights Act 1997)—to hold community meetings about the CCA. Based on the CCA factsheet, the council apparently agreed to "specific conditions for the protection of the forest and associated habitat and species within [the area] as part of the entire MMPL" (CI 2010 6). Farmers were deputized as "community rangers [who] agreed to help ensure that land and resources use within the area complies with the provisions of relevant laws such as the SEP [Strategic Environmental Plan]" (6). This meant that fellow farmers were tasked with monitoring and disciplining their neighbors; straight from the biopolitical playbook, Pala'wan rangers policed their own people's swidden and hunting practices.

CI also offered villagers incentives to align with the conservation agreement. At about eight hundred meters above sea level, staff targeted several large Pala'wan families who relied on a patchwork of swidden fields, fallows, and older forests that overlapped with the "no-touch" core zones. These families' main sources of subsistence and income—and the basis of their well-being—hinged on unfettered access to and use of these forests (e.g., swidden, NTFP harvesting, and daily wage labor, *arawan*). The CCA aimed to overcome such forest dependency through a combination of strictures and incentives:

> The agreement specifies conservation actions to be undertaken by the resource users, and benefits that will be provided in return for those actions. The conservation actions to be undertaken by the resource users are designed in response to the threat to biodiversity. The benefits are structured to offset the opportunity cost of conservation incurred by the resource users. The types of benefits vary, but may include technical assistance, support for social services, employment in resource protection, or direct cash payments. (CI factsheet, n.d., 1, cited in CI 2010)

CI hired Pala'wan to build a large house resembling a Pala'wan *kalang banwa* (big house) from forest materials. This served as the NGO's multipurpose tribal hall-cum-environmental learning center (see fig. 6.4). Here, Pastor Pilmar would teach Pala'wan elementary English and Filipino, educate them on the importance of conserving forest flora and fauna, and instruct them to adopt alternative livelihoods and reduce their reliance on forest resources. The center itself was equipped with a chalkboard and images of bountiful watersheds and colorful forest animals. CI's educational reforms cut across all age groups, with children and village leaders alike enrolled in the alternative learning center. Twenty-seven

FIGURE 6.4. Tribal Hall, alternative learning center, Marenshewan, Southern Palawan, 2013. Photo: Dressler.

enrollees, aged between five and fifty, attended lessons on Mondays and Tuesdays. The CCA Progress Report (CI 2010) relayed that most Pala'wan enrollees learned how to use a "pencil and ballpen" and how "to identify colors, shapes, simple numbers; [and] memoris[ed] the Filipino alphabet" (7). All these educational tasks were usually reserved for the provincial school system.

The CCA intervention also offered a range of material and financial incentives—packaged as "cobenefits"—to urge the Pala'wan to embrace sedentary livelihoods and do more with fewer forest resources. These cobenefits included the distribution of pigs and chickens, one year of financial support for an informal education "para-teacher," school supplies in the learning center, and support for coconut planting, corn production, and vegetable gardening. One year of financial incentives was also extended to the community to facilitate and sustain forest patrolling activities. This involved fifteen days of patrols by two Pala'wan "forest rangers" paid at ₱1,000.00 (US$20) per month for one year. New Pala'wan rangers—most of whom were swidden farmers—were then selected after two weeks. CI also had Pala'wan establish and manage a nursery of Indigenous tree species to help reforest at least ten hectares of denuded upland areas each year (CI 2010, 7). The introduction of cash crops, tree crops, environmental education, forest patrols and monitoring, paid reforestation efforts, and agroforestry plots all aimed to facilitate fixed-plot, sedentary agriculture and curb movement into older forests farther upland. CI's reform agenda was direct and multifaceted.

CI's CCA Progress Report (2009–10) claimed that the CCA program was broadly successful in optimizing Pala'wan behavior, livelihood, and bodies across the categories of income, education, health, and cultural integrity. The average annual community household income had supposedly increased by 42 percent, excluding payments for planting trees and monitoring the area. The report also claimed that leprosy, malaria, and respiratory and waterborne diseases were all in decline. Because of this, community members seemingly valued the CCA's support and had not violated the agreement in twelve months. CI considered its biopolitical program a resounding success, with the Progress Report (CI 2010, 14) concluding that "the conservation agreement scheme was proven effective in terms of forgoing destructive activities by directly addressing the socio-economic needs of the resource users. The communities have right away received and felt the benefits they identified, thus, they also adhered to the necessary conservation actions to protect the nearby old-growth forest and rehabilitate damaged patches." The report went on to claim that "there were no incidents of hunting, timber cutting and opening of new kaingin area. . . . Some 500 hectares of primary forest classified as core zone [were] maintained" (CI 2010, 13). However, the discussions I had with several Pala'wan involved in the project suggested otherwise. Most had become despondent and frustrated about the CCA initiative. Few were hopeful about any immediate or future benefits, and most continued to clear swiddens in mature forests.

Contrasting Realities

Sitting in his modest home near the national highway sometime in 2013, Pastor Pilmar initially affirmed that the CCA program was achieving its objectives. After talking about the program's design, he explained enthusiastically:

> Well, for the two years of teaching them, so far it's good, and they [Pala'wan] like it, in our literacy program they already know how to write and to sign their signature. In terms of environmental teachings, they have also learned a lot. That's why as you can see, it's been two years now that there is no kaingin [swidden] here; they have really stopped [clearing the forest].
>
> So, I teach them basic literacy, but I am mandated [by CI] to include environmental awareness in my lesson plan. I tell them they should protect the forest, that they should not cut trees beside the river, and that they should not kill wildlife!

However, when I later asked about other aspects of the CCA's implementation in Marenshewan, the pastor conveyed mounting frustrations:

> Well, according to the agreement between CI and the tribe, this tribe here was initially supposed to be given ₱2000 monthly, with two [Pala'wan] forest rangers each receiving ₱1000 monthly [US$22]. The idea was to change [rotate] the forest rangers monthly. That was supposed to continue throughout the year.
>
> With that money, they would guard the forest, they would act as forest rangers, and plant seedlings in kaingin . . . but this only happened a few times. Now there's nothing. It was to be over a year.

After contacting CI and the local government to track down the payments, he suggested that the Pala'wan "forest rangers" could be repaid in bagged rice rather than cash payments. Unfortunately, neither request was honored. He noted that, "in the end, we agreed that instead of giving them money, we would convert it to a monthly supply of rice. But that also never happened."

The pastor's own sense of "pastoral care" for the Pala'wan of Marenshewan, some of whom now regularly attended his branch of the Assemblies of God Church, was not being fulfilled as he had hoped through forest conservation agreements. Our lengthy conversations about the plight of katutubo and the Pala'wan of Marenshewan made it clear that he believed only Jesus could help them overcome their "pagan" forest-destroying ways. Despite these misgivings, CI had found their ideal biopolitical broker for what closely resembled evangelical conservation.

Panglima Bilog's Fatigue and Frustrations

CI's main tribal broker and tingkep trader, panglima Bilog, was also experiencing project fatigue and frustration. During one of our first conversations, he described how the arrival of the NGO had dissuaded him from relocating his community from the less productive lands of Marenshewan to a flatter, more fertile area of old growth forest at lower elevations. Bilog had hoped that relocating downstream would allow each family to cultivate about seven hectares of this productive land and improve upon their rice yields. He described the situation with a hint of desperation:

> We really wanted to relocate to the old growth forest to clear about seven hectares of kaingin. It's flatter and fatter land to do kaingin. There's a river there that is good source of water. The reason we cannot open that place is because CI will not approve of it. That's why we are asking other

people; maybe you can help us, Doc? According to them [CI], we are not allowed to open an area if it is giba [old growth].

I then asked if CI was offering the community any alternatives through the CCA. Bilog responded somewhat defiantly: "That's why I would like you, Doc, to document this situation. If they will not live up to the [CCA] contract, I will get out of it, and we all go back to kaingin. We've waited for nothing." A few days later, my conversation with Bilog about the CCA continued but in a more frustrated yet defiant tone:

> You know, Doc, the reason why we can't stop doing kaingin is because it is part of our culture. It is the main livelihood we have. There are times when we have rice, and times when we have no rice, but we still survive because we plant cassava. When we do kaingin in bunglay [secondary growth] it will not yield a good harvest; we only get small root crops, but if plant in giba [old growth] we can get big root crops. But now we're told to avoid making kaingin in giba!
>
> Right now, there are people here who explain [to us] about Mount Mantalingahan and that we should protect and preserve it; we are told to avoid making kaingin [swidden] in giba so that the watershed will not be damaged.
>
> If they want us to stop doing kaingin, we need an alternative livelihood that will not make it difficult for us to live. Now, they built that house here and a nursery but the income from that is still small; we only get a bit of money if we plant seedling, but none when we stop planting. They told us that if we do not destroy any areas with big trees we will get a total of seventy chickens . . .
>
> Yes, this is all from the office of CI and that's the agency that forbids katutubo [Indigenous peoples] from doing kaingin. If we cannot do kaingin, we have nothing.
>
> So right now, we just obey them. . . . Sometimes I think that it was much better before CI came. Before, when we had kaingin, it allowed us to survive; if we don't have kaingin, I feel that we've lost everything. We had big hopes, but now there's nothing.

Another Pala'wan farmer from the same village area agreed with his panglima's sentiment, claiming,

> CI is the agency that helped us so that we will not cultivate the giba [old growth] and avoid its destruction. Because that's their appeal—they don't want many of us to open new giba for kaingin. But in this area,

I'm not responsible for making kaingin. I also follow others to not open [forest for] new kaingin. But we were promised livestock and funds for forest protection, but none of these came through. They promised the farmers here more money if we plant [trees], but until now nothing has happened.

These Pala'wan farmers echo the resentment of many Indigenous highlanders who have experienced a litany of nonstate project interventions that overpromise and underdeliver across Southeast Asia (Asiyanbi and Massarella 2020; Dressler 2017; Massarella et al. 2018). They highlight the negative social and ecological implications of tree planting initiatives in swidden fallows. For them and CI, tree planting was a political act with contrasting interpretations and consequences; many farmers were frustrated that the trees they had planted for the CI project made it increasingly difficult to clear and burn, while CI believed they had curbed swidden-induced deforestation. Ultimately, a fundamental discrepancy existed between CI and Pala'wan understandings and expectations regarding the CCA's implementation, particularly in terms of livelihoods. This resulted in perverse ecological outcomes: by "fixing" fallows with newly planted tree crops, swidden clearings are displaced elsewhere, often into old growth forests. As Pala'wan farmer Diego Darmon explained:

> Before CI, our livelihood was kaingin but when they came, we couldn't do it anymore. They prohibited us and we are told to guard the Mount Mantalingahan including the virgin forest—they told us not to touch it. But then they told us that if we plant seedlings in our kaingin we will get money. But we thought that if we plant the trees in *dati kaingin* [older, reduced fallow kaingin] we cannot go back there [to clear in bunglay, secondary forest] . . . so we have problems because it is illegal to open new kaingin in giba. I told them [CI] that if we plant fruit trees in our bunglay [fallowed secondary growth kaingin], where will we do our kaingin now? And they said that we can use other bunglay, but we still cannot open a new one. When we only clear and stay in bunglay we cannot produce a good harvest . . . the soils become weaker [*mahina ang lupa*]. So, we have no choice but to open older forest.

Another farmer described the CI project and its tree planting initiative with considerable ambivalence and a hint of sarcasm:

> I can't disagree and I cannot agree. . . . We just follow along. . . . I mean, we received chickens; one lived, the rest died because of a stomach

infestation. So they say, "If you don't cut down the old forest, we will give you a chicken."

But we were also told to plant a lot of trees in our uma [newly cleared] fields. I was given four hundred saplings and was supposed to be paid four pesos per tree. We planted many of them in our fields but after a while we had to stop, and then planted all the trees in one area only [outside of the uma]!

Our panglima suggested we do this. . . . He said that if we plant all these trees in every kaingin, in the future we won't have any area for kaingin. We really don't have a choice *but* to open in giba.

Indeed, panglima Bilog's position held firm, noting that he and his fellow farmers planted most of the trees in one specific area so as not to interfere with clearing and planting in swiddens within old growth. He exclaimed: "We will not plant it all, because if we do, where will our kaingin be now? We have to open giba, so we still have a place for our kaingin!"

Ultimately, CI's two-year campaign to have Pala'wan farmers replant and monitor forests through cash incentives and financing led to the opposite outcome: the opening of older forests for swidden and resistance to the project. In cases where farmers *did* plant tree crops in fallows, they typically cleared vegetation around "high-value" saplings (e.g., mango or jackfruit) for fear that burning a fallow plot would destroy their tree crops. Instead of clearing, burning, and planting what would otherwise have been mature fallows (with productive soils and fewer weeds), Pala'wan farmers were often compelled to clear and burn older forests further inland that overlapped with Environmentally Critical Areas Network (ECAN) core zones for better yields. If fallow areas devoted to tree crops expanded, less land would be available for proper fallow regeneration, leading farmers to mine soils and degrade forests.

Decades of NGO-driven afforestation initiatives—known today as "nature-based solutions"—have effectively constrained the forest areas available to clear, burn, and fallow on ancestral lands. As Novellino (2007, 199) found for the Batak in central Palawan, when uplanders are compelled to repeatedly farm *dati* kaingin (previously cleared, reduced [two-to-three-year] fallow swiddens), their fields become *maniwang* (thin) and "infertile, with poor yields and some fields producing less than 400 kilograms [of rice] per hectare." Historically, a long-fallowed field (ten years or more) would yield more than 1,000 kilograms of upland rice per hectare (McDermott 2000; Dressler 2014). Afforestation-induced intensification and sedentarization leads to lower rice yields and higher food insecurity over time (contra "enriched fallows"). Social relations also shift toward more individuated claims to fields as notional property, as happens closer to the lowlands.

Most farmers eventually stopped planting "CI trees" because of the piecemeal, inconsistent nature of the payments. The remaining saplings were simply planted in a single area or left to die in the nursery. Lantana vines had subsumed parts of the alternative learning center, and the front section is now used as a cooking shelter by Pala'wan hunters passing by on their way home. Most farmers continued to clear swidden in upland forests, as they had done for centuries.

Still, lingering effects of the CCA intervention remain. Pala'wan are typically cautious about where and when they clear forest for swidden and most worry about the repercussions of doing so. Many have also become deeply despondent about the false hopes and expectations that NGOs stoke (Dressler 2014), particularly those with novel "community-based" agreements.

Weaving in the "Sticky Tingkep"

The PTFPP and CI interventions did leave some "cobenefit" residue, albeit with varied consequences. Although the financial incentives did little to maintain the tree planting and patrolling campaign, those Pala'wan who were encouraged to weave more, better-quality tingkep continued to do so as a source of income.

CI enhanced the local tingkep supply chain by intensifying the weaving, supply, and demand of baskets with regular purchases and deliveries to souvenir shops in the city. The alternative learning center served as a main weaving hub, with baskets on display or hanging from ceiling rafters for eventual sale. A CI staff member brought the baskets from the uplands to direct buyers and souvenir shops in PPC, with the NGO claiming that "the community was able to sell 1,000 tingkep, 600 placemats, and 70 *biday* [hammocks] . . . with a total accumulated income of ₱32,000 (US$711) as of June 2010" (CCA 2010, 10).

As in Kamantian, the Pala'wan in Marenshewan soon began crafting and selling the baskets much more intensively. Seeing a potential financial opportunity, panglima Bilog created (like Sofrano) a tingkep monopoly, taking over sales facilitation from CI and then trading the baskets himself on behalf of village weavers and those living further inland. He typically collected baskets from households, stored them in his house, and intermittently delivered them to souvenir shops in PPC. Bilog noted enthusiastically: "If I can get more rattan, we can make more handicrafts and sell them in the market to make more. If I can produce more, I'll be able to sell them directly in Puerto and then have a bigger income." Other local panglima tried emulating Bilog's tingkep marketeering, working as both fixers and traders, collecting and selling countless baskets to shops in the city. Many, like Bilog, advised some families that it was better to sell their baskets to them rather than sell directly at markets.

Nonstate efforts further intensified handicraft production—from tingkep to wooden masks to brooms—fostering expansive supply chains running from the forested highlands to the rural lowlands and the city. NGOs leveraged technical interventions and the tingkep's cultural symbolism to facilitate trade from Brooke's Point to PPC, Manila, and beyond. In time, more Pala'wan uplanders became firmly integrated into these emerging trade routes but typically on uneven terms.

Leaving the Uplands, Selling Crafts in the City

Pala'wan shared similar experiences of the long journey from the uplands to urban centers to sell their crafts. Many traveled alone, negotiated with shop owners, and resided in boardinghouses, experiencing racialization, market entanglements, and immiserations in the urban realm. The biopolitics of humiliation and low prices in the city were enmeshed with reform efforts in the distant forests of the Mount Mantalingahan Range.

Direna Calcon

Every few weeks, Direna Calcon, a widowed Pala'wan woman from Kamantian, collects various-sized tingkep and carved wooden masks from Pala'wan villages high in the mountains of Brooke's Point and Bataraza. With her young Pala'wan helpers, she descends to the lowlands to embark upon the long journey to sell the goods in PPC. Waking before dawn, Direna skips coffee and packs her food for the journey, walking two hours along narrow forest trails with a dimly lit torch. She carries crafts, root crops, and chayote (*Sechium edule,* initially planted by the PTFPP) in a rattan pack (kiba) to sell along the way. The group finally reaches the coastal plains at daybreak. They walk along the bunds of paddy fields in migrant settlements to reach the national highway, the north-south highway connecting southern municipalities to the island's more urbanized, populous northern region. They walk along the highway's side-berm until a jeepney stops to take them on the two-hour journey north. The driver gives them a free ride but charges them for the "handicraft cargo" (two pesos per item).

The group disembarks at a stop near the city center before the jeepney reaches the New Public Market, the main market outside the city center. Over the next several days, they will hawk their wares on the city streets, moving from one souvenir shop to another. Direna narrates their travails: "We came here to make a living from the mountains. We left early yesterday morning. We walked down from the mountains at night and then walked along the highway for two kilometers. . . .

We took our products on the jeep to sell . . . masks and baskets from many different families!"

The already difficult day's work in the city is made more uncomfortable amid the hot sun, dust, and diesel fumes of "Puerto," as the locals call it. Nights are spent on the cold concrete floors of boardinghouses or in the back rooms of migrant-owned souvenir shops. Pala'wan hawkers sell as much as they can to their preferred buyers or whoever wants their crafts (e.g., tourists lingering in front of hotels or waiting for shuttle buses to take them on "ecotourism" trips or to the airport). Some sales involve haggling or may not be realized at all.

Shop owners who have an oversupply of baskets and other crafts in their warehouses are often reluctant buyers, interested only in the best wares. Others may only purchase "traditional antiques" and high-quality crafts from well-known carvers. Both scenarios are problematic: the valued customary possessions of uplanders, their material culture, can be quite literally "emptied out" from forest homes as objects and heirlooms for ritual and ceremony (e.g., brass gongs, blowguns, and older tingkep) are bought or traded away from Pala'wan families and their communities (Johnston 2014).

My own visits to the backroom warehouses of older souvenir shops revealed what appeared to be dozens upon dozens of old gongs, spears, pig jaws, winnows, mortars and pestles, Pala'wan musical instruments (e.g., kudlung) and, of course, tingkep piled in disuse and disrepair (see fig. 6.5). While shop owners source these items from upland villages on the basis of their presumed authenticity and exotic character and value, their poor treatment in the warehouses betrays a disregard and contempt for Pala'wan culture. Unsold goods are returned to depots or to the families who made them in the uplands.

Malwang Vicente

Malwang Vicente from Marenshewan has been a trader and hawker for at least a decade. He travels between the uplands and PPC, with stops in Bataraza and Brooke's Point, to sell handicrafts from Marenshewan, Bono Bono, and other upland villages. Like Direna, he wakes up at 3 a.m. and, by 4 a.m., Malwang, his wife, and his child are carrying their load to the national highway in search of a shuttle van or jeepney. He is less fortunate than Direna and pays for his travel and the cargo, which eats into any profit he makes in the city: "I pay ₱200 for one jeepney trip, and then ₱50 per sack of cargo. One time, I had four sacks [bundles] of baskets, which cost me a total of ₱200."

Upon arriving in the city, Malwang visits the souvenir shops and, if sales are poor, proceeds to open-air restaurants and hotels and hawks to tourists passing by: "I pass as many souvenir shops as possible. . . . Those who want to buy will

FIGURE 6.5. Unsold Indigenous objects pile up in the dusty backroom of a souvenir shop, 2013. Photo: Dressler.

buy. But I keep touring around until I sell as much of my products as possible." I recently learned that Malwang has scaled up his operations; he now plies PPC's streets on a hired "trike" (a type of tuk-tuk) to sell his goods to larger department stores (e.g., NCCC, Alala Store, and Budget) and the pharmacy Drugman. Most store owners buy small brooms and masks: "If they have fewer stocks for sure they will buy, but if they have plenty of stocks, they won't."

As a trader, Malwang keeps the difference between what he pays Pala'wan weavers and what the shop owners pay him (see fig. 6.6). This is seldom enough to break even after all costs have been considered. He explains:

> I have been doing this for eight years now already. I do it because it gives me a small income and helps with the livelihoods of other families living in Marenshewan. Sometimes I collect baskets from other villages far away, such as Morin and Kulisian [Rizal]. I only buy from the people who make the best baskets in the community. . . . If there are other "rejects" I don't bring them here, because they don't sell.
>
> On good days I make the most money selling the smaller baskets to shops. . . . I am told that the tourists find them easy to pack. . . . The

FIGURE 6.6. Malwang selling his people's tingkep, 2013. Photo: Dressler.

larger ones take up more space. But many times, when I bring a delivery to the shops, they may not have the money to buy them all or they may have too many, so I will return the baskets to those who make them. I go back empty handed.

Not only is the experience of traveling from the forested uplands of Kamantian and surrounding areas to the busy city costly, time consuming, and subject to low returns, but many Pala'wan traders and sellers are also racially discriminated against by urban dwellers and souvenir shop owners. Malwang and others detailed how city dwellers ridicule them as "tribal," even while they purchased their "authentic" tribal handicrafts. Malwang recounted one incident in particular:[6] "Yes, some laugh at me. . . . One buyer told his companion, 'Ah, look, that is a tribal [katutubo].' But his companion replied, 'Don't tease him. He has plenty of money because he's so industrious in selling his products compared to others who just wait for help.'"

As this quote suggests, Pala'wan are perceived as "lazy tribals" until they sell their products and emerge as "industrious natives": closer to being Filipino, or equal marketized citizens. Such racialized perceptions of "entrepreneurial natives" as productive, self-sufficient market subjects are echoed in the biopolitical practices of both the PTFPP and CI (Li 2018).

Souvenir Shops, Discrimination, and Exoticization

In the late 1990s, migrant-settlers from Cebu, Luzon, and Mindoro opened several souvenir shops, trading under names such as "Tribal Handicrafts," "Yunika," "Nativo," and (perhaps less troubling, but equally curious) "Asiano" and "Manunga"—an appropriated Pala'wan term ironically used in reference to *taw manunga*, meaning "do-gooders" (Revel et al. 1998). Some of the first souvenir shops were established by the children of pioneering parents who migrated to Palawan in the 1950s in search of land. These migrants' initial dispossessory land claims and capital supported their entrepreneurial activities in the city several decades later. In two cases, the parents of larger shop owners had claimed sizable tracts of land (approximately five hundred hectares) in the south, near Bataraza and Rizal, upon which upland Pala'wan worked as day laborers.

The uneven agrarian political economy in the south partly facilitated the accumulation and concentration of capital in urban areas farther north. The wealth generated from claiming and exploiting Pala'wan land and concurrent sourcing of Pala'wan carvings and baskets allowed migrants and their offspring to invest in and establish their shops in the city. As one shop owner noted:

> My parents traveled to Bataraza a long time ago to purchase about five hundred hectares of lands for paddy rice, copra production and banana plantations [fifty-one hectares were privately titled] near the barrios down there. This land was flat and rolling and was already being cultivated by the natives. But we had a case to remove the [Pala'wan] squatters on this land. I then managed about fifty hectares of all of this land in 1978 and I then transferred to Barangay San Ignacio. In San Ignacio, I downsized and only had three hectares of paddy rice and five hectares of kalamansi and banana in the uplands.
>
> Things got tough so I decided to sell orchids and parrots and other birds in Puerto but the DENR [Department of Environment and Natural Resources] cracked down on me . . . so eventually I sold all my lands. So, after seeing that a lot of them [Pala'wan] had things, and after I sold my land, I decided to open this souvenir shop.
>
> It was the first shop in 1985 and eventually there were about six stores who only sold handicrafts made by natives, but from other tribes such as Igorot and not just those here. When we started out the shop was much smaller. . . . Before this shop was only about one-quarter of the original size.

Most shop owners held little regard for Pala'wan land rights and even less regard for their suppliers. While only a few came from families who had seized

Pala'wan land, most spoke disparagingly about the forest-dwelling Pala'wan, the tingkep, and customary handicrafts in general. For them, craft sales were just a way to make money. Most discussions eventually turned to disparaging comments about tingkep quality, the supply chain difficulties caused by "lazy natives," and tourists' broader interests (or lack thereof) in the basket's provenance. One shop owner on Rizal Avenue stated disparagingly: "If they have food, they stop making baskets. If they have hunger and expenses, then they make baskets. We only get half our orders. But this reflects the original Pala'wan way of life, and not the civilized Christian way—but . . . we still value the traditional aspects of the basket."

In contrast, the owner of "Yunika" ("Unique Objects"), Mary-Beth—hailing from Mindoro and having arrived on the island in 2005—began buying Pala'wan crafts in earnest a few years ago. In a conversation about her sales, she noted that only a few tourists ever asked about the baskets' origins, and that she did not really know (or care) where the handicrafts came from. Despite this disinterest, she nonetheless leveraged the crafts' "native character" to boost sales: "I don't know what the baskets mean [culturally] and only know they come from somewhere in the south of the island. . . . Most of the tourists are only interested in the native crafts, so we sell them that way, saying they're from 'katatubo.' Tourists aren't really interested in crafts made by people from around here. The curious tourists, though, ask, 'Who made this?' . . . 'Did it really come from native people?' We say, 'Yes, of course. They put rice and clothes in the basket to stop the rats from eating them.'"

In the context of this broader discrimination against katutubo in the Philippines, the PTFPP and CI's converging biopolitical work further reinforced efforts to reform Pala'wan existence across rural and urban areas. As Li (2007, 67) notes, nonstate actors hope that, by changing and optimizing the conditions under which uplanders live, a new set of practices and desires that align with outsider aims will emerge. Converging nonstate interventions have worked through the important material cultures (particularly the tingkep) long embedded in Pala'wan ways of life to facilitate reforms by weaving together limitations, constraints, and potential from the uplands to the city. The experiences of Pala'wan tingkep traders revealed the contingent and often fraught nature of these newer market engagements and livelihood practices, the outcomes of which produced uneven and unreliable returns for both traders and producers and further racialized uplanders.

While it is hard to say how extensively these interventions nudged Pala'wan social relations, labor time, and sense of property toward a narrower, individuated realm, those who were incorporated into marketized practices of reform—craft supply chains, fixed farming schemes, sanctions, and incentives—would certainly reflect on the ways in which markets and capital influenced how they made a living in upland forest settings. In this respect, the forging of environmental subjects was only ever partially realized and required additional reforms. However, this may not have mattered significantly. As chapter 7 demonstrates, the AFM's deeper program of pastoral care, extending from Kamantian deep into the Mount Mantalingahan Range, addressed any unfinished business.

Nonstate agendas and practices of reform have produced governable social spaces and political territories at the fringes of state jurisdiction and control in southern Palawan (Eilenberg 2014). This chapter has shown how NGO logic, infrastructure, and practice intersect to discipline and optimize the livelihoods, behaviors, and beliefs of the Pala'wan. Rather than governing human behavior through sustained subjugation (Foucault 2003), the PTFPP—working as an unadulterated nonstate entity—CI, and later the AFM, aimed to influence the free will and behaviors of upland subjects with a mix of incentives, improvements, and sanctions to align desires and aspirations with specific constraints and possibilities (Li 2007). These reforms were polyvalent and worked through individuals, families, and communities in spatially bounded conservation territories, aiding in coordinating access and use restrictions. Building on state environmental zoning and property rights systems, several decades of nonstate interventions have reproduced a complex spatial matrix of livelihood programs, environmental education, infrastructure promises, and, more recently, direct payment schemes aimed at discouraging the Pala'wan from engaging in extensive forest uses and upland living in favor of intensified lowland agriculture, market relations, and individual property rights. This, they claim, is the basis of sustainability.

The tingkep fetish was central to such reforms. Actively working through lowland traders and Pala'wan brokers, the PTFPP and CI intensified tingkep production, strengthened the supply chain, and enhanced market sales to provide upland families with the additional income that they hoped would somehow offset the need to clear and burn older forests for swidden and dissuade them from hunting game. Tingkep sales offered an important and culturally relevant source of income. However, the intensification of craft production resulted in a range of unintended consequences, such as resource degradation, indebtedness, and opportunity costs for other activities. Young women wove tingkep late into the night under solar lamps (self-purchased and NGO-donated), while their male counterparts marketed the products to souvenir shops and tourists in less familiar cities.

Along these commodity circuits, the tingkep's character was further commodified and partly alienated from familial relations, ritual practice, and livelihood across upland and lowland spaces. To sell tingkep and other crafts in city markets,

Pala'wan traveled the long and arduous route from the uplands to hostile urban settings. The tingkep and other customary items were slowly drawn from familial relations, ritual practice, and livelihoods into the commodity relations of lowland markets and uneven political economies. Chapter 7 considers how the AFM evangelicals applied similar but much deeper logics of reform—an amplified logic of benevolent care—to render the Pala'wan as Christian subjects, an effort slowed and complicated by customary practices.

7

FOR THE SAKE OF GODS

These former Taw't Deram [Pala'wan] cannibals are finding Jesus.

—Adventist leader Kamantian

The Adventist leader's evocative quote encapsulates the ideological foundation and profound outcomes of sustained nonstate reforms in the Kamantian highlands. Missionaries like the Adventist Frontier Missions (AFM) build on histories of state and nonstate civilizing imperatives to invoke a biopolitics of "benevolent care" to reach the "unreached" in the uplands of Palawan (Foucault 2007, 127). This is an evangelical biopolitics that works relentlessly from remote outposts. It aims to forge faithful Christian subjects through the intimate, personal spheres of upland life to establish deeper personal relations between Indigenous disciples and God. This, Adventists hope, will loosen the grip of adat and "the devil" and open uplander hearts and minds to their faith (Li 2007, 67). Missionaries work carefully. They study Indigenous languages, social relations, and forest settings to learn how they might make pious subjects of individuals, families, and larger communities. They patiently and persistently extend pastoral care through "zeal, devotion, and endless application" of ideals and strictures on Indigenous bodies and thoughts (Foucault 2007, 127). In some instances, they avoid maligning the free will of individuals to seduce, recruit, and reform more disciples; in other instances, they sustain reforms by shaming and denigrating custom among the most respected of Indigenous leaders. These are all calculated practices.

This chapter describes AFM biopolitics in practice and how the Pala'wan have responded socially and materially to deepening pastoral care that reforms through Pala'wan bodies, beliefs, and practices. It describes how AFM biopolitics operate from the inside out to instill a moral narrative of religious purity and modernity within and between families, along kin lines, and in local leadership

structures. Proselytization interventions constrain the use of rituals and livelihoods involving customary objects like the tingkep; in Keane's (2007, 6) words,
"in this narrative, progress is not only a matter of improvements in technology,
economic well-being, or health but is also, and perhaps above all, about human
emancipation and self-mastery." For the AFM, the pursuit of moral progress
requires that the Pala'wan completely detach from any "material mediations" of
spiritual "premodern" things and the constraints these impose over the autonomy of their agency (7). This separation of customary items and practices from
everyday life is the AFM's overriding biopolitical imperative.

Evangelical interventions built on the governance architecture of the Palawan
Tropical Forest Protection Program (PTFPP) to pursue extraordinarily visceral
substitutions of customary practice with Adventism through incentives, strictures, and sanctions across facets of the public and private realms of upland forest
settings (e.g., traditional healing and mortuary practices, hygiene, diet, livelihood,
and social play). I examine these dynamics of substitution and erasure in the contexts of reforming beljan relations with their own and nonhuman worlds as well as
disciplining livelihoods and ritual through expanding pastoral power. I describe
how evangelical practices were wielded through both the bodily and sensorial
dimensions of Pala'wan lives and broader community controls through the surveillance and procurement of sociobiological processes, and how the Pala'wan
negotiated these influences through alignment, resignation, and resistance.

Yet this missionary pastoral care has less to do with coercive force than "networks of material, social, technological, and technocratic relations [that induce]
the violence caused by pastoral forms of power designed to caretake" (Guarasci
and Kim 2022, x.). In many ways, evangelical biopolitics runs through Pala'wan
society as a form of cloaked benevolence that enacts violence within and through
Indigenous social relations and reproduction. The violence of pastoral care and
power tends to be less overtly socially and physically conflictual; rather, it is
manipulated and engendered as regularized social practice that is both implicit
and explicit in character. As Dumont (1992, 149) notes, violence can emerge
as "social practice"—a type of "habitus" that is "structured and structuring" in
what might be called evangelical "sacrifice zones" (Lerner 2012): heavily modi-
fied spaces wherein residents are subject to penetrating sociocultural alterations
to emerge as pious subjects in ancestral forests. These are not Scott's (2009, 7)
upland "shatter zones . . . where the human shards of state formation and rivalry
accumulated willy nilly, creating regions of bewildering ethnic and linguistic
complexity." Instead, they emerge as spaces in which the sacrifice of the body
and soul for Jesus is the order of the day.

The AFM in Kamantian hoped to enact a profound and penetrating reform of
Pala'wan families, cultural figures, and leaders. Upland shepherds actively steered

their Pala'wan flocks away from customary objects, social relations, healing, and play through self-discipline, incentives, and strictures. Central to this was the pastor and his disciples slowly redefining the customary roles and relationships of respected beljan (so-called witch doctors), important objects like the tingkep, and customary healing practices. Rather than admonishing the Pala'wan for the pragmatic use or sale of certain objects (e.g., baskets), the AFM sought to prohibit the use of customary objects in rituals and other forms of social reproduction connected to nonhuman worlds. They restricted the elaborate Tingkep 't Kundu ritual that used tingkep, gongs, kudlung, and the beljan's melodic incantations to invoke the female spirit of the forest, Linamen.[1] They actively disparaged and ridiculed beljan healing practices, ontologies of death, and convivial play among the young. And they worked across Kamantian and other villages deep within the Mount Mantalingahan Range to purge and purify Pala'wan customs—the social, spiritual, bodily, and material—in order to render older forms of social reproduction obsolete. The AFM's work would only be complete once customary beliefs and practices were fully reconstituted into Christian ones.

Drawing on ethnographic vignettes and the Kamantian AFM's (publicly available) online material, I reveal how mission staff and leaders reflect on their persistence, successes, and failures in proselytizing the Pala'wan toward religious enlightenment (Dunch 2002). The sect enacted manipulative and teleological narratives of reform, often by denigrating beljan to convert other family members and suppressing customary practice. While certain Pala'wan resented the AFM's call for moral, modern realignment, others engaged with the mission on a pragmatic level. Many Pala'wan in Kamantian proper also became deeply entangled in Adventist doctrine and reforms. In such conjunctures, the risks to customary life and livelihoods were dramatically amplified.

Optimizing Social Bodies and Unmaking Tingkep

The evocative quote that opens this chapter—"These former Taw't Deram [Pala'wan] cannibals are finding Jesus" (Anonymous 2016)—is the closing sentence of an Adventist leader's blog describing the AFM's desire to penetrate deeper into Pala'wan territory. The AFM aimed to reach and convert the Taw't Deram in Kebgen, a Pala'wan subgroup vilified as pagan and cannibalistic. Achieving this involved a Pala'wan convert, AFM doctor, and religious leader descending out of their helicopter to administer medicine, instill God's wrath, and overcome Satan—the AFM proxy for adat.

As the PTFPP neared completion in 2002, the Seventh-day Adventist (SDA) pastor and his wife took over their bunkhouse, clinic, and school for their own mission. The SDA already had its own literacy and religious program in place well before the PTFPP's arrival and harnessed other biopolitical practices to further weave evangelical aspirations into Kamantian and beyond. In time, the pastor and his wife had developed a primary school, a high school, a Sabbath school, a church, a small hospital, living quarters, and a (now operable) helicopter with a well-maintained landing strip.[2] The missionaries' living quarters were of better quality than the *kubos* of their Pala'wan disciples, which flanked the AFM staff quarters. During my first visit, I lived in the middle of this disciplinary matrix. Each morning, at around 5 a.m., I awoke to a speaker broadcasting a sermon about the right way to live: the Adventist way.

In 2013, while the pastor and his wife were on sabbatical in the United States, several young American missionaries managed the infrastructure and initiatives. Rather than supporting the construction of a (secular) high school sponsored by the provincial government (after the PTFPP learning center closed), the AFM instead built their own Adventist secondary school, elementary school, and dormitory for Pala'wan girls to live in during the week. As of 2014, the elementary school held about sixty-five Pala'wan students, while the dorm held four or five Pala'wan girls. There was also a cafeteria where Pala'wan cooked food for young Pala'wan disciples and other mission staff members. All the schools were said to be at capacity and in need of expansion to accommodate a growing number of disciples (AFM 2014a). These schools fused Adventist faith and formal schooling; reading, writing, and math were taught religiously to enable moral progression.

The bodies and beliefs of the young and old were also subject to evangelized biomedicine through "free" health checks and treatments at the main clinic and in the interior. The American AFM nurses used their clinic, knowledge, and beliefs to blend evangelism and modern medicine into a potent cocktail of proselytization. They efficiently and effectively treated illnesses long diagnosed and managed differently by the beljan's customary healing practices (and unsuccessfully, according to some Pala'wan). By medically treating the Pala'wan in ways the beljan could not, the AFM strategically secured their goodwill, trust, and allyship while undermining the legitimacy of the beljan's customary authority and power to heal physically and spiritually. The AFM knew that saving Pala'wan children with antibiotics and other treatments could help warm resistant Pala'wan parents and elders to the AFM community, Baptist beliefs, and "redemption through Christ." In doing so, the mission hoped Empu Banar would become less relevant and ultimately obsolete. A promotional video filmed at Kamantian and featuring the AFM president details the breadth and depth of the treatments offered at the clinic:

> The clinic starts tomorrow morning and they have about five thousand patients enrolled. There's no diagnostic equipment; here there's no other medical care, so whatever walks to the door you must treat it.

If you're a second year, third year nursing students, you may have to be delivering breech birth, performing a caesarean section on a wooden table with almost no equipment. So, if you're a missionary with the FM here in the highlands and you're in the medical profession or the nursing profession, you have an incredible opportunity to minister to people. . . . The healing ministry of Jesus is extended here to the highlands. (AFM 2014a)

Each of these five thousand Pala'wan patients had their biometric details (e.g., estimated age, sex, height, weight, ailment, area of residence) documented and incorporated into the mission's system of surveillance, monitoring, and control. The records were not limited to Kamantian but also included villages across the Mount Mantalingahan Range. Medical records in hand, the AFM trustees had considerable power to make their Pala'wan flock more legible and governable, reforming them into docile, quasi-citizens subject to de facto governmental interventions. By registering and cataloging the intimate social and biological details of Pala'wan individuals and families, the mission knew more about the biomedical history and details of Pala'wan—and had far greater control over such records—than most state-funded provincial hospitals in the lowlands.

In 2013, a group of young missionaries reflected on how their treatment of Pala'wan who had contracted cholera had brought the "pagans" closer to God. During my stay at that time, several Pala'wan confirmed their people had died from an illness unknown to the resident beljan. The AFM clinic staff understood this illness as cholera originating from water contaminated by a decomposing body close to a stream used by households along the outer edges of Kamantian. Based on the AFM's 2012 online account of the situation, it becomes clear that the mission understood the clinic's work as an opportunity to convert the Pala'wan by treating them and steering them away from customary healing practices:

It all started when Lebas died in the village of Lap Lap. Fifteen years ago, Lebas fell from a tree, and his health had declined ever since then. Before he died, he asked to be put in a tree house instead of being buried. Animist Palawanos worry what will happen if their bodies are buried and dirt touches them.

A few days after Lebas's death, two children from his family died. Suspecting that Lebas's body was a source of contamination, we told the village they should bury it, but no one was willing to do it because they were afraid of his spirit. In fact, most of the village had already moved to another part of the mountain out of fear. A few days later, a woman was carried to our clinic in a basket. She was severely dehydrated and barely responsive. Soon after, two more people were carried in. Fear was written across all the villagers' faces. We sent out word that anyone who was feeling even a little sick should come and get treatment. We also sent two missionaries to properly dispose of Lebas body and educate the village on how to prevent the spread of disease.

We ended up treating 16 people for what appeared to be cholera. Ten of them were in-patients in our four-bed clinic, so there were people sleeping on the floor and on the front and back porches. After two days of treatment most of them felt much better. Before we let them go home, we had them bathe and wash all their clothes. It turned into a big social event as each person took their turn. One older man we treated was fiercely independent and quite stubborn throughout his treatment. When he was feeling better, he told us, if it wasn't for your clinic, there wouldn't be anyone left in our village, just chickens.

The end of the story reveals how the young AFM staff celebrated their biopolitical moment. Not only did they treat the cholera, but their treatment also brought once reluctant and stubborn Pala'wan elders closer to "God's watchful eye," further from beljan healers and nonhuman worlds:

> We are so grateful to have been given the opportunity to help in this clinic. In the past [they were not] receptive to the clinic or the church, but this cholera epidemic has opened a door. . . . Now a group of high school students has been doing outreach there every Sabbath afternoon, and it has been well received by the elders and villagers.
>
> The past few Sundays for market day, our former patients from Lap Lap have been first to arrive, beaming with smiles. Out of a bad situation, God has provided an opportunity to read the village . . . for eternity. (AFM 2012)

The AFM's reductionistic interpretation and treatment of illness and death collides with the ontologically fluid understandings of illness and death among the Pala'wan. Most Pala'wan believe that illness is caused by "moments and outcomes of a complex network of interactions, involving not only society, ancestors, 'masters of animal game,' 'master of illnesses,' but also the whole universe and people's present and past" (Novellino 2001, 83). Pala'wan ontologies of illness, healing, and nonhuman worlds reflect an understanding of their own transition from an age (or celestial space) of benevolent deities (diwata), immortality, and abundance to the corporeal realm involving a time of relative scarcity and sickness caused by prohibited human acts disrespectful of the spirit world (e.g., incest, *sumbang*) (Macdonald 2007; Novellino 2001, 79; Smith 2021).

Pala'wan mythology describes a time when the Pala'wan still lived in the same universal sphere as God, Empu Banar, the creator who brought all creatures into existence. Here, agricultural surplus and forest foods were easily sourced; plants and animals were in abundance and gave themselves up easily. New conditions of illness, decline, and environmental extremes (e.g., flooding and drought) were brought about by unreasonable or sinful acts that ruptured the broadly "symmetrical" relations between humans, forest environments, and deities (Novellino 2001, 79–82). In the present human realm, the Pala'wan must contend with the material and immaterial responses of benevolent and malevolent deities responding to their behaviors and practices, amid Adventist missionaries attempting to strip these beliefs bare. As Novellino (2001, 82) notes: "Only from this moment on, humans acquire a perishable body with multiple 'souls': a life force or intangible essence (*kurudwa*) entering and filling the body through the whorl of the hair (the region of the fontanel), and other minor kurudwa located on the knees, elbows, and ankles. Furthermore, the acquisition of 'magic' (*pangti*), 'healing dances' (*tarek*) and other shamanic [beljan] practices are the immediate effects of the separation of humans from benevolent deities."

Beljan ritual specialists continue to matter in Kamantian, even as the AFM aims to make them redundant. They mediate the interface of the corporeal and spiritual worlds to contend with illness, healing, and hunting. Respected beljan, typically men, or the all-seeing Tungkol (a most powerful beljan) inherit and learn (from their fathers and or in situ) complex knowledges and practices associated with the healing properties of plants, animals, and rituals. Other Pala'wan also learn how to respectfully hunt, collect forest foods, and cultivate certain plants (e.g., rice) endowed with kurudwa to avoid unsettling malevolent entities (Dressler et al. 2018). As chapter 2 discussed, Empu Banar assigns specific divine "gamekeepers" to animals that can cause illness when certain rules of conduct are not respected—for example, *ungsud* offerings (Macdonald 2007).

The Pala'wan appease capricious nonhuman entities through ritual offerings or exchange practices to offset complex illnesses emerging at the edges of human-nonhuman worlds. Some use mutja and other charms to ward off disease, while others distance themselves or avoid certain areas and actions. Many heed the signs and sounds of deities in dreams or interpret certain forest species as warnings from malevolent spirits or a beckoning from benevolent ones, who, in both instances, may occupy the species (Novellino 2001, 83–86; Theriault 2017).

Death for the Pala'wan involves "elaborate ritual precautions" (Macdonald 2007, 104) that draw on the beljan's ritual work and require the attention of family members. Immediately after a death, family members keep vigil over the corpse to ensure that no evil spirits (e.g., lenggam) enter the body; this period of calm respect settles the deceased and insatiable spirits. Daily livelihood activities slow considerably, since unsettling malevolent entities may inflict undesirable outcomes on humans and the environment, including rice crop failures (Macdonald 2007, 105–6). While mortuary practices vary, the body of the deceased is then generally wrapped in bamboo lattice and buried in a secondary growth forest (never in old growth forests) (Macdonald 2007, 105–6). The body rests facing upward toward the rising sun, accompanied by important belongings such as baskets (open tabig or closed tingkep), machetes (*tukaw*), pots, and cloths covered by layers of bamboo, grasses, and soil.

The beljan then initiates the keluwatan ritual to "separate the dead from his/ her spouse and the rest of the living" (Macdonald 2007, 105). At this time, a final meal is prepared to appease the departed, mark their separation from the corporeal world, and avoid incurring the anger of the deceased's kurudwa, which lingers near the body and may threaten living family members (105). The satiated kurudwa may then travel to the house of Empu or to the land of the dead (*kelebegang*).[3] Generally, the beljan warns others to avoid contact with the souls of the dead and avoid offerings of food, whether in dreams or otherwise. This cautious respect for death and dying is crucial for facilitating the departure of the kurudwa into the spirit world (Macdonald 2007, 107).

In other cases, I was told that families must maintain a respectful distance from the corpse of someone who fell ill and died from encounters with invisible, malicious forest-dwelling spirits, lest they disturb the departed kurudwa (or contract the same ailment). Unintentionally disturbing *Satjan* in their abodes— in forests, fields or home—accidentally brushing their long, gnarly hair as they roam, or behaving disrespectfully (e.g., talking or speaking in excess or slander) can lead them to strike and bite as well as dislodge and capture one's kurudwa, leading to illness and possibly death (Revel 1990, 101).

Many elders describe how disrespectful encounters can lead to the feverish illness, *tulpok*, from touching or brushing forest plants, wood, or other objects that the demons have contaminated with disease (*sakit't marahan, sakit't satjan*). It is a broader category of disease best dealt with by beljan. He first attempts to treat tulpok-related illness by using the same plant or object thought to have caused the illness, with the characteristics of the illness and healing object being analogous. In contrast, others speak of the atmospheric ailment(s), *kadadak*, derived from weather (the winds of *habagat*), cool air, or the breath of evil spirits that cause chills, fevers, nausea, stomach aches, diarrhea, and liver (*atej*) sickness (*lagbay*). Different plants with traits reflecting the pain of kadadak are rendered into drink for treatment (e.g., near Rizal, sharp durian thorns are grated down and mixed with water, their sharpness being similar to the sharp stomach pains of kadadak).

At times when initial interventions fail and diseases intensify, it is likely that satjan have captured or claimed one's kurudwa. Only a beljan can ensure its release. In a trance, the beljan communicates with satjan and/or a higher deity, such as Empu't Parey, for precise diagnoses and to determine the type and quantity of offering needed to appease satjan and release the kurudwa. If properly appeased, and a deal is struck, the kurudwa returns to the body, removing the illness. If no deal is struck, the kurudwa is likely lost, and death is near. A beljan's clairvoyance is crucial for restoring a respectful "symmetry" between living humans, their health, and the (needs of) nonhuman realm. An unsettled kurudwa (and associated misdeeds) suggests poor weather; illness and bad omens will linger. If the AFM overcomes the beljan's restorative powers across body, soul, nature, and spirit worlds, they ultimately succeed in rupturing Pala'wan worldviews and intergenerational transmissions of oral knowledge and practices of healing.

Converting Healers and Nonhuman Worlds

The AFM and other missionaries have closely studied the beljan's principal and authoritative role in mediating and maintaining symmetry across the human and nonhuman worlds of Pala'wan highlanders. Several AFM blogs and my own observations suggest that SDA religious leaders and staff strategically target beljan by suppressing and delegitimizing their customary authority and power in the eyes of the community. In doing so, they work incrementally to disrupt the pivotal role the beljan play in fostering certainty, community, and confidence in the social life and relations of Pala'wan. Apparently, when a "witch doctor" is reformed, other Pala'wan follow suit.

In a blog post entitled "He Fancies Himself a Witch Doctor," an AFM religious leader narrates with a violent mixture of disdain, pity, and hope how the tenacity and authority of "witch doctors" may be overcome through persistent, close-quarter proselytization (Anonymous 2018). In targeting the few remaining beljan, they hope to dismantle the religious specialists' (and thus community's) cultural firewall against Adventism so other families can be brought into the mission's pastoral fold. Juxtaposing the power of Christ and SDA science against the beljan's "pagan" practices and mannerisms, an AFM leader condescendingly writes:

> Once upon a time in a faraway jungle lived a small, wiry man with darting eyes and wandering speech. He fancied himself a witchdoctor. Whenever someone was sick, he would appeal to the evil spirits to remove the sickness. Or if it was a difficult birthing situation, he would

appeal to the dead ancestors to accept the gift of a blanket (after the birth they would take it back and use it again) and allow the baby to be born.

When the missionaries arrived, the witchdoctor made friends in a groveling sort of way, looking for opportunities to get something he wanted from them. Whenever a patient requested help from the missionary clinic, he would lurk around waiting for a chance to do his "medicine" when the missionaries' backs were turned. And he wasn't the only one. Other witchdoctors would do the same. They took their work very seriously, especially if someone was possessed by an evil spirit. They were certain that only their craft would free the person from their condition.

The witchdoctor was married and had many children. His wife was very quiet, intensely shy and remarkably lonely. As the children grew, their opinion of their father diminished. He was an angry man, lazy and often cruel. His children wanted desperately to attend the mission school, but he would discourage them, even ridiculing and threatening them. Still, many of them persevered. One even graduated from the elementary school, and several were baptized.

This family lived just across a creek from the missionaries, and the missionaries heard lots of arguments and crying babies, but also children sweetly singing the songs of Jesus the older siblings had heard in school and were obviously teaching to the younger ones. What a contrast!

In the beginning, the young wife would make excuses to go and visit the missionary lady who would tell her Bible stories. As she came more and more frequently, a villager commented that she was taking Bible studies. "No, I'm not!" she retorted, and she began visiting much less often. Saddened, the missionary continued to pray for her friend. She spoke to her about spiritual things as there was opportunity, but when the husband was there, he would answer for her that she wasn't interested.

One day the wife told the missionary that she had been talking with her older children, and she really wanted to learn about God, no matter what her husband said. Would the missionary please come to her hut and teach her on a regular basis?

And so regular visits began. Progress was slow. . . . But she loved to learn and sing songs of Jesus. Her first favorite was "The B-I-B-L-E" in her native tongue. The studies were slow, but her interest and sincerity were real, so the missionary kept going and praying, amazed at the obvious changes the Lord was working in the young wife's heart.

One day when it was time for the usual study, the witchdoctor was there, but he said he wanted to hear the story, too! That was an answer to prayer. But it still wasn't smooth sailing. He would challenge the Bible stories and tell his own versions mixed up with his spiritual beliefs and ancestral legends. Praying for wisdom and tact, the missionary persevered. For several years now, the missionary has been working with this family—thrilling at any progress and sorrowing at the man's sporadic signs of disinterest.

The story depicts the reluctance of beljan and other Pala'wan to give up their customary roles, forest medicines, and rituals, while the AFM leadership persists with deep evangelical reforms amid the familial tensions they created. Many Pala'wan perceive a profound ambivalence and tension between the persistent proselytization of the missionary leaders, their customary beliefs, and their respected beljan who have managed to (re)affirm Pala'wan existence for centuries. The AFM denigrates the dignity and legitimacy of these revered cultural brokers, who have long cultivated and transmitted knowledge and practices to restore certainty concerning health, livelihood, and beliefs by nourishing and appeasing the needs of the spirit world in times of troubled relations. The Adventists' resolve to undo the beljan's customary role is tantamount to violently dismantling Pala'wan social reproduction, material relations, and identity. Their end goal is cultural erasure. Without the beljan mediating and recounting the meaning of the spirit world, the multiple worlds Pala'wan must navigate are lost in translation through predatory evangelism and conservation reforms.

Discipling Livelihoods

The AFM camp also established its own livelihood-trading system connecting labor production in the uplands and lowlands. Pala'wan disciples worked in SDA-owned paddy rice fields in lowland areas and wage labor jobs in Kamantian proper. Their daily wage was ₱100 per day—less than the then-standard ₱200 (including lunch) for manual labor—and they allegedly bought rice from the missionary paddy fields in which they labored (at prices higher than lowland rice granaries). Under the pastoral gaze of the mission leaders, other Pala'wan performed carpentry work on camp housing and infrastructure and assisted with maintaining the Sabbath school and church. Meanwhile, young American AFM volunteers undertook the core evangelical labor of the missionary complex: the teaching, nursing, automotive repair, carpentry, and audiovisual work needed to inculcate and broadcast their biopolitical doctrine.

In the state's absence, these governmental practices brought AFM staff, Pala'wan, and Adventism closer together. The foreign AFM staff and Pala'wan disciples worked together to ensure that the trading system, enrollment in Adventist institutions (Sabbath school), and mission-centered play (singing) achieved the appropriate balance between monitoring and strictures to support more profound proselytization. In this way, the AFM aimed to create an autarkic evangelical ecosystem that would sustain moral reforms and potentially be replicated by Pala'wan disciples across time and space.

In addition to these evangelical structures, the AFM leaders drew on SDA religious doctrine to establish local strictures to reform both Pala'wan body and soul. The mission first sought to narrow their diverse forest-based diet by exhorting them to avoid "unclean" meat, which would morally pollute them (thereby complicating their religious freedom and ascent to heaven). These "polluted" meats included omnivorous hoofed species, such as the wild Palawan bearded pig, *biek talun*, (*Sus ahoenobarbus*): a spiritually important and frequently hunted species that supplies the starch-heavy Pala'wan diet with crucial calories, fats, and protein. Some Pala'wan also believed that the consumption of marine and riverine species without scales and fins (e.g., crabs, shrimps, clams) was prohibited. Despite this, many families—particularly women and young children—actively harvested freshwater shrimps (e.g., *urang*), freshwater crabs (*kayangat*), and river snails (*susu uyaw*) by hand from numerous fast-flowing rivers and pools. Larger, plump snails (e.g., *patung balay*) and other finless and scaleless species were also collected from moister forests and fallows.

Adventist religious dietary strictures also prohibited Pala'wan from using blowpipes (*sapukan*) and thorny rattan rakes to catch any avifauna (scavengers, detritivores) that consumes fresh meat or carrion, including owls (*gukguk*) and bats (*ememkung*). Prohibitions against consuming such forest and riverine species constrain hunting knowledge and practices, further complicating the cultural symbolism of the hunt and kill, particularly with tingkep et mutja.[4] As the Pala'wan typically hunt and collect varied game, use of the tingkep to carry smaller game—bats, birds, Palawan flying squirrels (*Hylopetes nigripes*)—with the powers of mutja gives them the advantage of greater strength and confidence during the kill. The alternative protein source of Adventism's "clean" split hoof, ruminant animals (goats, water buffalo, and sheep) are typically too costly for most Pala'wan to rear and maintain. While the Pala'wan do kill and consume such undulates, this rarely includes goats and draft animals, such as carabao, which are essential for plowing, hauling, and breeding.

The Tingkep 't Kundu Erasure?

The AFM also attempted to steer the Pala'wan away from customary rituals (e.g., Tingkep et Kundu), traditional healing practices (with *ruruku*), storytelling (*tuturan*) of myths, and the performance of epics (*tultul*). Many Pala'wan living in the vicinity of the main mission camp were dissuaded from using and exchanging customary material culture, including gongs (*basal*), ritual dancing (*tarek*), two-stringed lutes (*kudyapi*), and the bamboo zither (*pagang*) for the Tingkep 't Kundu ritual (see chap. 2). These rituals are intentionally conspicuous in their performance, with loud gong assemblages and ritual chanting (*deruhan*) facilitating the beljan's deep trance as he mediates between human and nonhuman worlds. He elicits answers from benevolent spirit beings (diwata) about everyday concerns and future events, including sickness, hunting potential, and changing weather (Macdonald 2007). However, the missionaries consider such gatherings, particularly with rhythmic chanting and gongs, the work of the devil. In Kamantian proper, these instruments rarely accompanied singing for courtship, ritual gatherings, or other forms of play. Similarly, few elders performed epics and storytelling in ways that could bring Pala'wan young and old together to listen to and circulate the meaning of legends and heroes central to their lifeways.

Many of the Pala'wan with whom I spoke were also alarmed by AFM prohibitions against smaller—but highly consequential—customary offerings such as ungsud (i.e., cooking lutlut before opening a swidden field for a good harvest or to appease lenggam in or near Lihien forests). As Ubud Marmang, a Pala'wan elder, noted: "We make lutlut but it is not good for them [the missionaries]. They don't even let us eat the lutlut because [the rice is mixed with] fish, and they say that fish with scales should not be eaten. We cannot even eat chicken.... It's prohibited.... So if we can't hunt near here, we get meat in the lowlands."

Ubud Marmang's assumption that the missionaries prohibit the consumption of scaly fish and chicken suggests that Adventist strictures are generally understood as broader prohibitions against the consumption of all "wild meats" from forests and rivers. Those who engage directly with the Adventists have a clearer understanding of the sect's dietary biopolitics. For example, Ramog Kulibit explains a slightly different situation: "One time a group of young Pala'wan who had been baptized were attending a worship session on a Saturday. After that they were prohibited from eating pork." When I asked how the AFM did this, he responded: "If we make offerings with tingkep, they tell us that our Gods are not the true Gods and that's why meat should not be eaten. But we are the same as they are. We call on our God, and they call on their God. We cannot stop the missionaries, but we still teach our young how to make lutlut. They will sustain the practice."

Rather than forbidding the local dialect, the AFM conducted church services and Bible readings in the Pala'wan language to ensure that they could sustain their reforms; apparently teaching Pala'wan English was actively avoided for fear of "outsiders" (e.g., anthropologists) making them aware of their rights. With the Bible translated into Pala'wan, this mission preached in Pala'wan dialect and trained Pala'wan pastors to finish what the Catholic Church had failed to do decades prior (Macdonald 2007, 4, 1992a).

Disciplining the Backslide

Breaching AFM religious strictures and norms resulted in various punitive sanctions for Pala'wan disciples. Although sanctions appeared to be applied intermittently, they amounted to comprehensive disciplinary tactics of control of uplander ways of life in and beyond Kamantian. Objects of play (e.g., volleyballs and basketballs) and customary materials needed for rituals were monitored and disciplined into disuse. The AFM ensured that, through sermons and word of mouth, the broader community was made aware of the strictures underpinning such disciplinary measures—and the consequences of breaching them. For example, any Pala'wan who missed church services were noted and fined (approximately twenty pesos) for "backsliding" from the faith: a term describing those who focus on meeting their daily needs and concerns through customary practices rather than consistently adhering to the Adventist faith. I had learned that even Sophia Tamaran, an elder caring for her husband, who had been paralyzed from a stroke years earlier, was charged with backsliding despite years of commitment to the mission. Apparently, her journey of spiritual emancipation through Christ had stalled due to providing daily care for her ailing husband.

The AFM often celebrated their successes in the "conduct of conduct." On an AFM blog, an Adventist leader describes the difficulty of "training and developing [Pala'wan religious] leaders" to a "higher level" at which they feel sufficiently comfortable to "pray publicly, share personal testimonies, call for the offering and read scripture aloud in church" (Anonymous 2021). When young male Pala'wan mission leaders flourish, it is supposedly because their conscience is no longer hostage to adat custom; in the Adventist's own words, successful reform is when they no longer fear "provid[ing] direction and correction to someone who is older or holds a higher position," such as an elder or beljan. In one case, a pastor describes how a young Pala'wan man—an Adventist "trainer"—was religiously maturing because he and others concluded that volleyball and basketball should be discontinued in view of the morally indefensible village pastime of betting on team sports. These fledgling disciples acknowledged that betting even small amounts of money or goods (e.g., bead necklaces or cigarettes) on the outcomes of sports (or cockfights) was unacceptable behavior for good Christians. To set a public example, they decided to cease playing volleyball altogether. The Adventist leader described the discussion on his blog:

> One day as the men were outlining what they felt were the most important standards to teach members to uphold in order to not be like the world, they listed 15 areas specific to Palawano culture. During our lunch break it suddenly dawned on us that they hadn't mentioned anything about gambling or betting. I was surprised, because this temptation has a long history among Palawanos.
>
> When the afternoon session opened and this topic was suggested, they all agreed. Then one of the trainers asked, "Historically, you, who were members in the early days, made the standard that Christians should not play volleyball, whether the game was being bet on or not, and whether or not the team members were fellow Christians."
>
> Jonan replied, "Yes, that was the way it was then. We were concerned that the temptation would be too strong. But now I tell my Bible students that it is okay to play volleyball as long as you don't bet on it. I love to play volleyball, but I don't bet anymore. It's okay."
>
> The trainer responded, "Isn't it a bit tricky when we start using ourselves as an example of what is right or wrong? Furthermore, how does an onlooker know that you are just playing volleyball and not gambling? Isn't this creating a temptation for those who are drawn to gambling? What is our responsibility as Christians?"
>
> After more prayer, a lengthy discussion ensued. The topic expanded to basketball, traditionally a non-betting game for the mountain Palawanos. The lay pastors admitted that there were now some groups that did bet on basketball as well. An older lay pastor ventured to ask, "What about the amount of time spent playing basketball? We are called to accountability for the use of our time, right?"
>
> After discussing back and forth, sometimes quoting scripture, sometimes telling stories and experiences, they finally took a vote: Is it okay to play volleyball or basketball? The vote was a unanimous "no." They decided it would be better not to be a stumbling block to others than to enjoy this pastime. Even Jonan agreed that he would give up playing volleyball and basketball, though betting was no longer a strong temptation to him. . . .
>
> I was amazed to see how these Palawano men had grappled with a very sensitive subject and addressed it biblically and prayerfully. In the end we all stood and sang a song of victory. I went home that evening amazed and rejoicing—both for our leadership at large and for the new attitude evident in Jonan. (Anonymous 2021)

Such an outcome reflected the AFM's persistence in applying pastoral care in the most public spaces of Pala'wan existence, where self-disciplining was enacted explicitly and performatively in front of others. In contrast to the assumed disciplinary success, however, most Pala'wan regard betting on village sports games and cockfighting (*tatabuqan*) as a long-standing convivial affair rather than a vice-laden activity that inevitably leads to social conflict and moral decay (Macdonald 2007, 45).[5] The act of betting reflects fraternal social play among Indigenous men from villages near and far, nourishing social relationships and identities. As Macdonald (2007, 57) also observed:

> Gambling and betting are also favorite pastimes, as is cockfighting, a sport practiced without the use of artificial spurs. The fight thus lasts a long time, and the defeated cock is not killed. Therein lies the main difference between the Palawan and the Christian lowlander's cock-fighting. Informants claim that the art of cockfighting among them is ancient and dates back from ancestral times. The marketplace provides a venue for encounters between boys and girls and facilitates discreet meetings between them. All this helps tighten the social fabric. (Davis 2013, 552)

Despite AFM strictures, Pala'wan betting and village social play persist, pigs (and other SDA-prohibited animals) are still hunted and eaten, and ritual ceremonies and customary practices involving Tingkep are still performed. In this sense, AFM reforms do not easily substitute Pala'wan customs—at least not in any lockstep manner. Many Pala'wan remain ambivalent, despondent, or resentful toward the AFM's (and the PTFPP's and Conservation International's [CI's]) incremental reforms working in upland communities.

Resisting Pastoral Care?

Many Pala'wan elders expressed deep concerns about the AFM's presence in their lives. The Kamantian residents I spoke with were divided over whether the Adventists had a rightful place in their upland lives and communities (Macdonald 2008). Broadly, those living near or in the AFM's nucleus appeared to be deeply devout (e.g., singing hymns and attending Sabbath school), while those living further upland often disapproved of the mission. Several Pala'wan families at the eastern edge of the community—about eight hundred meters above the mission camp—suggested that the Adventist faith was generally incompatible with their way of life. As elder Ramon Drago, who lived in the next valley, bluntly stated, "No. I will not join their church because they [the SDA] prevent us from eating what we want." This was echoed by the middle-aged farmer Maron Muntar, who added: "The missionaries have convinced many to join but have not yet convinced me or my friend, Baltazar. The others feel bad inside [upset] because they were taken and baptized."

Closer to Kamantian, Elder Rio Tulunag, who had long disagreed with the AFM's practices, noted: "They [the missionaries] forbid us from making offerings to our rice God. We are different. We do not understand their practices and there are prohibitions in them. The way they worship is different from us and we cannot understand their language. Our beliefs cannot be mixed with the missionary culture because they prohibit us from practicing our customs. Our culture is different from their culture; we cannot mix with them." Other Pala'wan went further. For example, Drago's next of kin and neighbor Sonja Bagrar decried the working conditions at the mission station, highlighting how their poorly compensated labor underpins its maintenance: "We do daily work in the missionaries' *basakan* [rice paddy]. The salary is not fair because it fluctuates daily, from 100 to 150 pesos. We then buy rice from the pastors' basakan at 90 pesos per *ganta* [which is normally around 40 pesos]. We also wash clothes for the missionaries. They pay too low." The low wages for manual labor were remarkably similar to the meager compensation provided to Pala'wan for the ecological labor they invested in afforestation work for CI and other nonstate actors.

Unlike the AFM's more devout disciples, many critical Pala'wan were not "true believers" (Paredes 2006, 522), keeping their family and homes away from the AFM outpost. Rather, they deflected SDA influence by embracing and defining customs, rituals, and practices. These Pala'wan continued to use the tingkep for hunting meat, ritual healing, and weaving knowledge and belonging across genders and generations.

Benevolent Care or Evangelical Sacrifice Zones?

The AFM leadership worked hard to ensure that their pastoral power was, as Foucault (2007, 127) described, "fundamentally a beneficial power ... to ensure [their Pala'wan] sheep do not suffer and do not stray off course"; such power reflected the shepherd's duty of care involving "zeal, devotion, and endless application." In recent years, the mission designed new plans to use its rendition of "benevolent care" to free the souls of pagans deeper within the Mount Mantalingahan Range. Somewhat sinisterly, the coupling of "liberating" scripture and biomedicine was once again being used to legitimatize and incentivize deep reform. In a 2014 online YouTube video, the lead pastor ruminates on extending the mission's pastoral care to the "untamed" Taw't Deram subgroup (AFM 2014b):

> If you look at the ridge up there at Mount Mantalingahan, on the other side is the Tagdaram territory. We have work started over there, it is

a twelve-hour hike from here. At the moment, there is nobody there. Logistically, we have not been able to keep anyone over there. We hope to reorganize our logistics to get some people over there to continue on the work. There's people who are ready, they are asking for schools. It is wide open. It is just a matter of getting the personnel and getting the logistics set up.

Evidently, the mission succeeded in reaching one of the Taw't Deram villages—by helicopter, no less—noting in a recent video that the (supposedly "previously unreached") villagers had now been "touched by Christ" (AFM 2019). A lead AFM pastor elaborates on the "benevolence" of their quest to reach and reform the Taw't Deram according to God's designs, noting:

> The believers in the *Kebgen* area are being persecuted for their beliefs— death threats, fines for leaving the belief system of the ancestors, and rejection from spouses—yet these people are joyfully staying true to the God they've come to trust. They have much yet to learn, and they still show many trappings of their former beliefs. Before, even with a smile on their faces, their eyes spoke fear and darkness; but now their eyes reveal their love of Jesus and trust in God. They are no longer controlled by fear of the supernatural, and they are willing to suffer derision and loss to hold onto their faith. It is a beautiful thing. (AFM 2019)[6]

Despite many Pala'wan being clearly troubled by the AFM's presence, the mission continued to proselytize under the guise of benevolent care, repeatedly imposing its religiosity and moral sense of right and wrong among a highland people known for their self-sufficiency, autonomy, and complex customary practices. Over time, the AFM's extended reach will have significant consequences for the sense of dignity, political autonomy, and self-determination of Pala'wan.

In these evangelical spaces, highlanders must negotiate multiple fields of biopower that invariably leave a mark upon their sense of personhood, community, and ways of life. In Kamantian, Pala'wan agency is thus forged through the everyday of upland living—and forms of resistance—amid the multiple interventions imposed by those who desperately seek to inculcate reform (Li 2007, 228). The AFM's omnipresent pastoral power and embeddedness ensure that their ideals, incentives, and sanctions interpolate Pala'wan lives in more ways than one.

AFM religious leaders built upon the existing PTFPP infrastructure to forge an evangelical theater and comprehensive biopolitical architecture: a clinic, schools, a church, dormitories, an airstrip, a helicopter, and financing have allowed foreign and Indigenous labor to maintain a self-sufficient mission with evangelical nodes that penetrate further into the interior to proselytize "pagan tribes."

A mixture of disdain, loathing, and pity drove the AFM's reforms ever deeper in southern Palawan.

The sociospatial logic of Adventist reforms in the highlands has reproduced what might be best described as an evangelical sacrifice zone: a distant, forgotten space in which customary practices—whether beliefs, livelihoods, or diets—are forfeited and renounced in favor of evangelism, purity, and progress (Lerner 2012). In these sacrifice zones, Adventists implement comprehensive and penetrating religious reforms that aim to reconstitute Pala'wan identity, beliefs, and practices across multiple realms of upland society (e.g., women and men, young and old, healers and leaders, satyan and Empu). The violence of these reforms usually occurs in the out-of-the-way spaces that avoid attention, where the rights of the Other are considered expendable, even when legible under Indigenous rights regimes (Lerner 2012; Farrier 2019). Pala'wan life is thus simplified and reduced to absolute categories of premodern primitives or enlightened Christian moderns, but seldom as full rights-bearing subjects. They bear the disproportionate burden of sacrifices that other "full citizens" can typically avoid (Lerner 2012). The outcome is an upland landscape sacrificed in the name of a Christian God (and conservation imperatives) in which people, livelihoods, and forests assume a narrower meaning and structure reproduced through profound sociocultural and biophysical rupture.

Building on decades of Baptist evangelism spreading across the Philippine highlands, the Adventists took on de facto government roles and responsibilities in the nonstate spaces of southern Palawan. With tacit approval from the state—and little, if any, oversight—the AFM worked with calculated precision and persistence in distant highland areas, patiently narrowing diets, prohibiting Indigenous rituals and associated material culture over many years. The AFM did not anticipate immediate results in its evangelical sacrifice zone. They worked slowly and comprehensively, violently mixing praise with punishment. In one breath, they lauded the religious work and devotion of Pala'wan disciples; in another, they publicly shamed and denigrated the legitimacy of beljan and chastised those whom they perceived as backsliding from Adventism. Building on their successes in Kamantian, the AFM strove to expand its political territory by replicating the conversion process deeper and deeper into the southern highlands. By installing devoted foreign and Pala'wan pastors in remote areas, they established makeshift churches, imparted the "word of God," and treated sick Pala'wan to entice them away from the adat of ancestors (or ancient ones, *kegungurangan*) and into the hands of Jesus. Seeing Satan in adat, the AFM labored to dismantle and reform a savage custom.

Multiple biopolitical constraints, pressures, and influences—sermons, strictures, new diets, hygiene, heaven, and hell—have all converged at the nexus of missionizing and upland living to forge enduring disciples at the margins of state rule in Palawan. AFM biopolitics have operated through body and soul to instil a moral narrative of religious purity and modernity. Here, the Adventists "wrestled [with] the role of material mediations in spiritual life" (Keane 2007, 6–7) in which the moral progress of Pala'wan was linked to their gradual detachment from any sociomaterial items and practices of customary significance (e.g., mutja, tingkep).

Such programs of reform are, of course, situated within and mediated through Pala'wan histories, sociomaterial relations, and aspirations, which complicate the rendering of uplanders as compliant subjects. The interwoven rituals and material cultures of the tingkep, other customary items, and swidden practices challenge nonstate actors' efforts to wrench apart Indigenous beliefs, livelihoods, and sociomaterial worlds. The human and nonhuman worlds of Pala'wan are still deeply layered, entangled, and remembered, often making "tribal purification" difficult, partial, and incomplete. Indeed, the Pala'wan engage nonstate actors and interventions in different capacities, with many aligning but others availing of project benefits with little, if any, enduring allegiance. Some highlanders have become disillusioned and despondent, while others resist such reforms more directly. However, as Li (2007, 228) observes, while "no one program fully shaped the highlanders according to the plan, . . . all of them left traces on livelihoods, landscapes, and ways of thinking."

CONCLUSION
Conjunctural Biopolitics

This book opened with the question of how and why diverse nonstate actors aim to reform the lives and livelihoods of Indigenous peoples in the highlands of the Philippines. What social ideals and economic logics motivate nongovernmental actors to deploy significant reserves of labor and funding to establish near-permanent programs and infrastructures—complete with technologies of rule, varied expertise, and codes of conduct—to overhaul and optimize the self-sufficient and sustainable existence of Indigenous highlanders?

For the Sake of Forests and Gods has shown how conservation and religious nonstate actors draw upon remarkably similar logics to patiently and persistently reform uplander existence for the ideological and moral ends of sustainability and consecration. In many ways, both are deeply faith-based interventions. After centuries of state attempts to govern highland areas, nonstate actors have reinvigorated ideologies and practices of reform to make Indigenous livelihoods, beliefs, and behaviors more legible and governable than any state actor ever could.

Often well-funded and organized, nonstate actors work incrementally and comprehensively to regulate social life and biological processes at the individual and population levels in seemingly remote areas (Lemke et al. 2011). Rather than employing coercion—at least overtly—most nongovernmental initiatives support and incentivize uplanders' free will, beliefs, and practices to optimize their social and physical capabilities, usefulness, and docility (Li 2007). Such reforms are both targeted and sweeping, loaded, as they are, with moral assumptions and values: mothers, fathers, children, healers and leaders are all nudged down the "righteous path" toward sustainability and Christian morality. Nonstate interventions potently intersect with Indigenous social relations and reproduction on a near-permanent basis, with enduring social and material implications.

Nonstate organizations have driven reforms through the everyday practices of social reproduction and, especially, through the sociomaterial things that matter most to Indigenous people in the uplands. Using the social relations and material objects that mediate social and biophysical reproduction—particularly the tingkep, associated rituals, and swidden—nonstate interventions render legible the social and livelihood attributes of Pala'wan supposedly most in need of reform. Deficiencies are identified and managed through a complex matrix of biopower: political knowledge, regulatory controls, monitoring, organizing, incentivization—punitive strictures (often cloaked with benevolence) that all converged to leverage, modify, and repurpose the dignity, beliefs, and subsistence of Pala'wan uplanders (Lemke et al. 2011).

Such reforms have involved community agreements, incentives, and strictures to align Pala'wan customs and livelihoods with both Adventism (i.e., outright bans on certain foods and suppressing rituals involving tingkep) and forest conservation (i.e., bans on swidden, planting more trees, incurring fines for clearing, and weaving tingkep for tourism). The nonstate panopticon extends well into the uplands to monitor and discipline Pala'wan social life and resource use. Yet this conduct is partial, resisted, or ignored as uplanders reappraise the potential and pitfalls of such interventions. Biopolitical frictions and everyday life rub against one another amid preexisting livelihood risks and uncertainties.

NGOs fill the governance voids left by state agencies' lack of political will, funding, or competence. However, the state is never entirely absent from such governmental rationalities. State and religious authorities have coproduced the uplands as marginal spaces in need of ordering and governance for centuries. They created the politico-legal foundation for deeper nonstate reforms in frontier areas. Nonstate actors are often characterized as cheap, efficient, and effective surrogates that expand colonial legacies and narratives of reform into the contemporary period (Clarke 1998, 2006).

I began this book with reflections on my initial fieldwork encounters with nonstate actors and their reforms in the southern highlands of Palawan. The introduction described how I came to recognize—off the back of twenty years of research in the uplands—that the Pala'wan faced pressing political and ecological influences from a motley crew of actors that were less concerned with state enforcement than environmental literacy and the second coming of Christ. In recent decades, diverse nonstate actors have moved into hinterland spaces to assume governance roles of reforming the Indigenous poor in line with an optimal, modern existence—one that approximates nonstate (and state) ideals of · Christianity, sedentarism, private property, and commerce (West 2006; Li 2007).

Chapter 1 used the notion of conjuncture to link the theory and practice of biopolitics, materiality, and the politics of difference. The biopolitics of conjuncture refers to how nonstate and state ideas, practices, and social relations "articulate together at a particular locus" (Massey 1994, 154) to influence how Indigenous peoples ought to make a living in and respond to the agrarian political economy of upland and lowland settings. While fragmented in application and reframed by Indigenous peoples themselves (Cepek 2011), I argued that the confluence of biopolitical ideals and practices undoubtedly affects how Indigenous uplanders perceive themselves, their lives, and their livelihoods. After expounding on Foucauldian biopolitics and pastoral power, I described how nonstate biopower can work through and powerfully reform some of the key facets of highlander existence—namely, social reproduction, livelihood practices, material culture, and indigeneity—across time and space. Linking theory and practice, I described how nonstate biopolitics and pastoral power attempt to forge compliant subjects by working through and altering the private and public spheres of these social registers of Indigenous existence.

Chapter 2 offered a window into Pala'wan life and livelihood in the southern highlands of Palawan. Drawing on my fieldwork as well as the foundational work of others, I described the Pala'wan as a semiautonomous polity characterized by complex social relations, livelihoods, and nonhuman realms that have allowed them to thrive in the uplands over centuries of colonial and postcolonial rule. I examined the central role that the swidden cycle plays in Pala'wan social and biophysical reproduction and how the persistence of swidden underpins the fabric of their worlds. I showed how the tingkep is forged from families, kin, and forest spaces that cut across human-nonhuman worlds to mediate everyday social relations and livelihood reforms. Nevertheless, nonstate actors aim to undo agroecologically diverse swidden cycles and forest fallows that nourish Pala'wan worlds.

Chapters 3 and 4 described the trajectory of colonial and postcolonial forest governance and missionizing practices. The overlapping Spanish and American colonial systems sought to eradicate swidden practices, instill ecological and hygienic behaviors, sedentarize, and invoke Christian ideals to expunge Satan from pagans in the uplands. These colonial ideals rendered Indigenous peoples legible through racialized codification and disciplinary practices to criminalize, fix in place, and assimilate complex livelihoods and cosmologies from the highlands into the lowland colonial administration. These intersecting state and religious governance ideals were carried into the postcolonial Philippines. After independence in 1946, nonstate actors took on state governance roles in distant forests. In the post-Marcos era, evangelicals and environmental NGOs moved into the uplands to establish a complex architecture of rule that (more than the state) worked through the public and private spheres of Indigenous existence. Reform was penetratingly intimate, and while punitive sanctions remained, new incentives and enhancements aimed to leverage expectations and opportunities to ensure seemingly willful enrollment and alignment. Conservation and Indigenous NGOs promoted new discourses of sustainability and Indigenous rights, fused with markers of indigeneity as the basis for reforming destructive livelihoods. Evangelicals like the Seventh-day Adventists approached customary ways, material cultures, diets, and nonhuman worlds with disdain and a desire to overcome them through an unrelenting reform of body and soul.

In Chapter 5, I examined how colonial ideals and practices have influenced contemporary nonstate governance of Indigenous social relations, livelihoods, and residence patterns in the uplands of Palawan. The significant Muslim Tausug presence and rugged mountain terrain limited the US colonial state and Catholic Church's influence on Pala'wan lifeways. The Pala'wan took advantage of a less accessible political space that gave them greater degrees of autonomy, unlike their Indigenous neighbors further north (Dressler 2009). The occupied and cultivated forested upland interior remained more or less beyond the reach of state foresters and the Catholic Church (unlike the lowland copra and paddy rice fields below). For the evangelical missionaries and environmental NGOs that emerged in the 1970s and early 1990s, these highland peoples and older forests seemed vulnerable, foreboding, and ungoverned—a "wild" space needing pastoral care. The Adventists and environmental NGOs exercised converging agendas for reforming the ideational and livelihood basis of Pala'wan ways of life. Whether expunging Satan or eradicating swidden burns, both groups of actors were (and continue to be) motivated by deep desires to overhaul Pala'wan existence in line with foreign ideals of social and ecological purity.

The state's political economy—the consolidation of property rights (private title and timberlands), marketized livelihoods, sustainability, and sedentarism—influenced nonstate governance motives, infrastructure, and interventions to coproduce and govern uplanders as modern subjects. This governance architecture aimed to control and flatten the human and nonhuman worlds of the Pala'wan. I described how important material cultures like tingkep became implicated in and mediated such practices of reform through family, ritual, and livelihood. Each nonstate actor's program and practices worked to enhance and control Pala'wan subjects by rendering legible markers of indigeneity, livelihood, and customs to reinforce their own and the state's work of governing upland peoples and places. Despite contrasting perspectives and motivations, these nonstate actors espoused similar disciplinary logics of optimization and reform.

Finally, chapters 6 and 7 described the contemporary biopolitical theater of nonstate reform and Indigenous encounters in southern Palawan. I described how nonstate governance reforms worked with calculated breadth and depth in the highlands. Environmental NGOs and the Adventist Frontier Missions (AFM) not only governed comprehensively—across individuals, families, and communities—but also focused on the social and material elements of cultural importance to the Pala'wan. They literally worked within, between, and across diverse spheres of social, cultural, and biophysical existence in the uplands and lowlands. These crafty shepherds knew that lasting influence depended on how well they monitored and disciplined aspects of upland life central to personhood and community cohesion (e.g., customary healers, objects, rituals, and livelihoods).

The Palawan Tropical Forest Protection Program (PTFPP) and Conservation International (CI) valorized Pala'wan custom and indigeneity (e.g., tingkep and other crafts) to supplement, substitute, and eradicate swidden-based livelihoods. Nonstate conservation actors worked hard to valorize the tingkep, turning a customary basket into a commodified livelihood supplement to curb swidden and reform upland living. Entrepreneurial Pala'wan became "craft brokers," buying and selling tingkep produced by other Pala'wan near and far in a supply chain stretching from the forested uplands to urban centers. This was, perhaps, a good source of income that potentially revitalized the craft. However, the production, sale, and branding of the basket and other customary objects further integrated Pala'wan into uneven market relations and the racialized imaginaries of souvenir shop owners and tourists in the lowlands.

Meanwhile, the AFM pushed a deeper pastoral care that, with a veneer of benevolence, persistently disparaged and contained Pala'wan customs, beliefs, and diets. AFM leaders were busy using evangelical tricks of the trade to dismantle beljan customary status, knowledge, and function in Kamantian and beyond. The AFM's religious rhetoric and evangelized science undermined the legitimacy of revered beljan and the powerful objects (e.g., mutja, ruruku) they use to manage illness and uncertainty by bringing families together. Such biopolitical violence reproduced sacrifice zones with enduring social and material harms.

Yet these violent biopolitical reforms are not fixed and do not produce absolute territories of control. While reform projects politically and sociospatially demarcate and delimit existence (e.g., zoning that spatially defines permissible livelihoods and residence), such spaces are indeterminate in character: porous, shifting, and contested by the Pala'wan and other actors (Yusoff 2017). Pala'wan took advantage of upland projects while fending off reform by ignoring, resisting, or exploiting opportunities and, ultimately, deferring to their ancestral ways, both in rhetoric and practice. In other instances, the Pala'wan drew upon their understandings of Indigenous rights—often framed and translated by state and nonstate actors—to better negotiate terms of engagement with overlapping governance agendas (Theriault 2017).

The Uplands as Spaces of Reform and Resistance

An increasing number of nonstate actors have classified, delimited, and reordered the uplands as sites of experimentation and reform. Since the 1980s and 1990s, nonstate reform programs have filled the void of the state with incremental and diverse interventions intended to reinforce categories of difference, ideological control, strictures, and incentives. Adventist missions and environmental NGOs gaze upward to discursively reframe ancestral lands as untamed and primitive spaces in need of control and reform (van Schendel and Maaker 2014; Eilenberg 2014). A range of nonstate actors now occupy, for great lengths of time, upland areas with people occupying forest landscapes—deemed illegible, deficient, and vulnerable—that must be improved. Technologies, conditions, and incentives overcome a pagan past to achieve an optimized, modern future (Paredes 2006; Li 2007). In time, unruly subjects emerge as the pseudocitizens of their nonstate trustees.

As studies of other highland peoples and places across the region have shown (Eilenberg 2014; Scott 2009; West 2006; Li 2007), NGOs, missionaries, and other nonstate actors have reimagined the uplands as "far from the historic reaches of capital [that] hold much of the in situ biological diversity" (West 2006, 5); as "untamed and wild resource frontiers" replete with high-value forest resources (Eilenberg 2014, 157); and as being filled with ungovernable, stateless peoples, who, despite evading state rule for centuries (Scott 2009), now require direction on how to become optimal natives.

Nonstate actors with seemingly contrasting objectives thus draw on similar biopolitical logics that fill government voids by fixating on upland life and livelihood (Foucault 1978). These nonstate actors attempt to restructure Indigenous social relations with newer regimes of authority that define "proper uses and users" (Rasmussen and Lund 2018, 388). Selling the legitimacy of such interventions is no easy task. As chapters 6 and 7 argued, whether enacted by an Adventist mission or an environmental NGO, successful reform requires that ideals and practices are worked through Indigenous identities, subjectivities, and customary practices on a regular and sustained basis across upland landscapes. These actors radiate out from their operational centers (e.g., Kamantian, Puerto Princesa City, or Washington, DC) to overcome an inherent fragility and vulnerability assigned to Indigenous lives and livelihoods in an untamed periphery that they themselves construct. Imagined binaries of the past—primitive, degrading pagan versus advanced, sustainable Christian—are revitalized to be overcome.

Nonstate actors pursue reform with tacit approval and legitimacy from Indigenous subjects. They leverage their benevolence and duty of care with "zeal, devotion, and endless application" (Foucault 1978, 127) through social contracts and incentives—punitively or otherwise—that define labor practices and value production. CI social contracts aimed to realign Indigenous ecological labor with markets, natural capital, and "sustainable" land uses (Joslin 2022). The Adventists established religious contractual ideals to forge docile, pious subjects (Semple 2017). The apparent benevolence underlying nonstate motives of betterment allows them to reproduce (somewhat coercive) project ideas and ideologies in the personal and intimate spaces of the uplands and influence how Indigenous peoples reproduce themselves socially, biophysically, and materially through new logics, ideas, and conditions. In other words, nonstate reforms chip away at Indigenous ways of life.

Although the PTFPP, CI, and SDA drew on ostensibly different governance agendas, their interventions intersected and aligned to reform Pala'wan ways of life toward a modern social order. Crucially, all nongovernmental players worked through key local brokers to launch and sustain their programs. The PTFPP and CI leveraged select panglima to further incorporate tingkep baskets into tourism supply chains, finding financial and social incentives for weavers to add value by making more, better-quality baskets. As NGOs often do, both organizations (and souvenir shop owners) marked and elevated the customary character of baskets in general and valorized the authenticity of smaller tingkep as ideal, portable Indigenous souvenirs. Sellers and buyers could easily identify the tingkep as a "truly" Indigenous Pala'wan artifact—an authentic heritage product. The income generated from such craft sales and livelihood support would supposedly offset the need to expand swidden beyond conservation zoning at higher elevations.

Mainstream conservation NGOs, such as CI, built on colonial state discourses to reinforce convictions that swidden agriculture is a destructive and degrading practice, despite decades of scholarly research showing its broadly sustainable character (Condominas 1977; Conklin 1957; Leach 1977). Across the region, NGOs work through respected Indigenous leaders to leverage their authority as project brokers and village mobilizers, ensuring that their subjects become compliant forest stewards whose ecological labor is used to monitor, guard, and plant tree crops in fallows to hinder further forest clearance and better align with program objectives and zoning.

Project Fatigue or Resistance?

While uplanders recognize that some social and material benefits have ensued from several decades of program interventions, they are also (justifiably) irritated by, and despondent about, the repeated requests to volunteer their labor and organize themselves for "community conservation" initiatives (Fletcher et al. 2019). Many highlanders would rather clear and burn their fallows than propagate and plant trees in swidden plots; they would rather eat or sell livestock instead of rearing it; and, instead of making crafts intensively for sale, most would rather produce them intermittently—and likely sell them at a better price—in line with other familial responsibilities and livelihood needs.

Even CI's pivot to direct payment schemes through Community Conservation Agreement (CCA) conditions failed to convince many Pala'wan to align with the program objectives of self-monitoring forest clearings, curbing swidden, and not hunting game in older forests. Many ignored or dumped trees meant to be planted in fallows. Elsewhere, Pala'wan have avoided giving project organizers too much information (see Rubis and Theriault 2020 for related examples) and expressed discontent about the lack of benefits and obstructions to making a living (Dressler et al. 2010). All of these actions reflect subtle practices of "noncompliance," "evasiveness," and "everyday acts of resistance" (Scott 1986; Kerkvliet 2009). Whether the Pala'wan considered these acts as explicit forms of resistance is, of course, debatable; many, it seems, are simply tired of the revolving door of environmental NGOs and false promises. Others simply "enlist" based on the shared expectations that they will eventually receive social and material benefits. Many leaders and villagers also pursue projects for political leverage and control (Titeca et al. 2020).

Entrustment and Coercion

Evangelical trustees are arguably more persuasive than other nonstate actors with whom they may travel. The AFM worked patiently and persistently in an attempt to condition Pala'wan as disciples through new ideals and incentives to purify body and soul. This patience, however, was often infused with a sense of urgency to overcome Pala'wan "cultural deficiencies" in preparation for the second coming of Christ. Like the PTFPP and CI, who enrolled trusted leaders such as panglima as project brokers, the AFM proselytized by selecting and converting Pala'wan who exhibited confidence and potential aptitude as religious leaders and local organizers. Other disciples were then recruited from within Pala'wan families and kin groups to help spread the Adventist faith within and beyond Kamantian. Rather than valorizing customary leaders (as other NGOs may do), however, this mission actively delegitimatized the beljan's cultural authority and varied roles in brokering well-being through customary healing practices and rituals that mediated uncertainty. The local process of enrollment and indoctrination was sustained through explicit, integrated incentives (health care and promotion in rank), conditionalities (dietary restrictions and labor in sedentary agriculture), and penalties (verbal reprimands, confiscating objects of play, and cash payments) that sought to discipline and destroy the vital role that healers, rituals, and customary objects play in sustaining Pala'wan ways of life.

The AFM facilitated this process of entrustment and coercion at an interpersonal level and in the public space of the community. The dynamics of Adventist enrollment and discipline cut across the private and public realms of Pala'wan life to deepen, broaden, and reinforce local recognition of and adherence to the mission's legitimacy and authority in Kamantian. Younger Pala'wan convinced themselves and their followers that various forms of play (e.g., basketball, volleyball, and cockfighting) induced a temptation to gamble and should be prohibited. This was both a deeply personal (shaming) and an intensely public (humiliating) approach to disciplining Pala'wan to align with Adventism. It was self-disciplining to the core. The very literal interpretation and associated transcendentalism of heaven (good) and hell (evil forces) reproduced a strict and bounded proselytization that, unlike indigenized Catholicism today, offered little maneuvering room for complex and changing Pala'wan customs (Macdonald 1992a, 135–36).

Affective entrustment and coercion involve what Semple (2017, 1–18) calls "turning strangers into kin." It is a relational dialectic that emerges through the intimacy of interpersonal relations between the governed and those who govern (Berlant 1998, 21). Here, entrustment and coercion can emerge through missionaries' construction of "lifelong affective networks" that incorporate Indigenous subjects into patterns and practices emblematic of the right Christian family and practice (Semple 2017, 1–18). Any form of trusteeship has the potential to supplant genuine citizenship and sovereign control, particularly in relation to the state's mandate to care for its citizens. Seduction, betterment, and sanctions foster trust and legitimacy in close-knit upland communities. The biopolitics of such pastoral care works extensively and endlessly to restructure Pala'wan social relations and reproduction across the uplands. It knows no end.

Vital Matter

Several Pala'wan families rejected AFM strictures by investing in and drawing on the social meaning and practices of customary objects and other material cultures vital to human and nonhuman relations. Most Pala'wan continued to hunt, using the tingkep to carry and eat prohibited forest game and carry powerful amulets (mutja, often pig's hair) to promote courage and strength. They continued to use the basket during extended rituals involving female forest deities and when storing divine rice for planting or collecting smaller fauna from fallows. The Pala'wan reaffirmed the legitimacy of their own social worlds through the Kundu 't Tingkep ritual. Like other customary objects in the highlands, the tingkep serves as a register—or, in Gell's (1998) words, an "index"—of social, cultural, and material attributes that influence inferences, responses, or interpretations between humans and nonhuman actors, and the social relations that are woven between them.

The tingkep's social and material attributes mediate, influence, and coconstitute social and material relations across the human and nonhuman worlds of Pala'wan. The tingkep, forests, and swiddens are woven together by "the Weaver," Nagsalad, reflecting "the visible knot which ties together an invisible skein of relations, fanning out into social space and social time" (Gell 1998, 62). The basket and its lively practices coconstitute the force of invisible spirit worlds, rituals, and ceremonies, which, weaving together life and livelihood, facilitate social reproduction among the Pala'wan amid intersecting biopolitical pressures (Bennet 2009). Weaving, using, and invoking the basket and other customary objects allow individuals and families to reaffirm ontological security in Indigenous materialities and partly refract nonstate biopower.

The Ungovernable Anticitizen?

Despite investing in customary practices and relations to contend with biopower, most Pala'wan have little to no recourse against nonstate interventions. Although the Pala'wan are born within the political architecture of the nation-state, most are not born as rights-bearers whereby "the rights of man and nation" are necessary and automatic (Agamben 1995, 75). Rather, highlanders occupy the liminal, distant realms located between state and nonstate spaces in which citizenship and the rights bestowed by the state are slow to emerge. Broadly removed from the state's sovereign reach and capacity to grant inalienable rights, Indigenous highlanders like the Pala'wan increasingly have de facto citizen rights bestowed upon them by nonstate actors hoping to optimize markers of indigeneity either for forest stewardship or Adventist discipleship (Chandler and Reid 2020; Greenleaf 2021; Neimark et al. 2020). Nonstate and parastate actors push modern ideals through new institutional arrangements, forms of cultural production, and the forging of subject rights in upland areas.

Nonstate actors legitimized their interventions beyond the provision of infrastructure and other forms of pastoral care neglected by the state. CI valorized certain customary practices over others (tingkep over swidden) to enroll, reward, and reframe Pala'wan as Indigenous forest guardians, not just degrading resource users. In contrast, the AFM's deep and varied interventions used knowledge of Pala'wan language and custom to ensure that their narratives of reform penetrated the public and private realm of Pala'wan lives. Where seemingly distant and punitive state reform programs fail, nonstate actors may succeed as the good shepherds of upland peoples through intimacy and proximity.

Nonstate actors have adopted the postcolonial state's ideological project of "civilizing the margins" through a combination of liberal and illiberal practices (e.g., Indigenous rights, marketization, sustainability, and evangelism; Duncan 2008). Various nonstate (and state) actors still perceive uplanders as the ultimate tribal "anticitizen," whose way of life and beliefs are vestiges of the past. In Lerner's (2012) phrasing, Indigenous highlanders have come to occupy "sacrifice zones" at the margins of state rule, wherein an influx of nonstate actors seeks to fragment and reform sovereign Indigenous polities. Nonstate actors pick up where governments leave off in reforming uplanders, who often simply retreat farther into the hinterlands. Well-funded and politically supported nonstate actors work zealously to reconfigure Indigenous existence and sociomaterial matter in line with perverse moralities and modernities.

This book describes a deeply violent biopolitics—one of erasure (e.g., of customary social relations, livelihoods, and reproduction) among otherwise self-sufficient and independent highland peoples. Questions remain as to how such biopolitical practices might be better negotiated and/or further resisted by Indigenous uplanders themselves. Should the state insert itself more prominently in the uplands to reaffirm its ostensibly more secular "social contract" with its most marginalized citizens? This could correct the corrosiveness of nonstate actors or simply reinforce the injustices of sustained, overlapping interventions. State agencies (and bi/multilaterals) will likely continue to outsource the governing and reforming of highland communities in Southeast Asia to politically aligned and effective nonstate actors. Such a foreboding future urgently requires that uplanders be more politically empowered to better negotiate the litany of biopolitical interventions coming their way.

Notes

INTRODUCTION

1. See also "Poaching Is Sending This Shy, Elusive Pangolin to Its Doom," *National Geographic*, May 14, 2019, https://www.nationalgeographic.com/magazine/article/pangolins-poached-for-scales-used-in-chinese-medicine; N. Unlay, "A Yearning to Return: For Travelers Living All around the World, There's Nothing Like Coming Home to the Philippines," *National Geographic*, October 21, 2022, https://www.nationalgeographic.com/travel/article/paid-content-a-yearning-to-return.

2. A *panglima* is considered an elderly customary leader, arbiter, or judge of senior status in the community. They are recognized experts in customary law. See C. Macdonald, Pala'wan Online Dictionary, accessed July 24, 2024, https://philippines.sil.org/resources/works_in_progress/palawan/dictionary).

As per protocol, I introduced myself and my research intentions and sought permission to conduct my research in the weeks ahead. Permission was granted in both study areas.

3. Most upland families across the island were navigating deepening market relations and resource use restrictions by reestablishing older livelihood options and or finding new sources of income. In addition to declining swidden rice yields, many uplander livelihoods are beset with lower supplies of overexploited (larger-sized) rattan stocks (*Calamus spp*); the tapping of heavy, low-return almaciga resin (*Bagtikbagtik*, manila copal, *Agathis philippinensis*) farther in the interior; variable, seasonal, and climate-specific harvests of moderately priced honey; and trying to secure on-farm labor opportunities in the lowlands. Despite many families still skillfully diversifying livelihoods to spread risk and ensure adequate (protein and carbohydrate-based) food provisions, those at the margins of, or more fully integrated in, market relations must engage with the rising precarity of upland living on Palawan.

4. The terms "uplands"/"uplanders" and "highlands"/"highlanders" are used interchangeably throughout this book to denote Indigenous peoples living in typically remote, hill, or mountainous regions in the Philippines and elsewhere.

5. The government's Provincial Environment and Natural Resource Offices (PENRO) is based far away in Puerto Princesa City, and local community, or CENRO, officers from lowland Brookes Point seldom traveled beyond the midlands (around four hundred meters above sea level). Despite the government occasionally granting Indigenous highlanders "native title" with fewer access and use restrictions, it was non-state actors who furnished ancestral domain titles through state laws and conditions (McKay 2006; Li 2000).

6. *Sukutan* is also referred to as a unit of heavier weight (twenty kilograms). C. Macdonald, Pala'wan Online Dictionary, accessed July 24, 2024, https://philippines.sil.org/resources/works_in_progress/palawan/dictionary.

7. The survey had several limitations. Not only was it intrusive and time-consuming, but the results were only indicative of general livelihood patterns and trends among the Pala'wan in both areas, lowlanders, the AFM, and commercial buyers. However, the survey did reveal the estimated volume of baskets being produced, where and to whom they

were sold, and for how much, relative to the income from other livelihood activities amid ongoing program interventions.

8. The pastor, his wife, and I (and most of the NGOs) were all professional strangers to the Pala'wan host culture. While each actor's position and objectives differed considerably, all were outsiders studying Indigenous host cultures in ways that generated friction and varied outcomes. One might assume that, with some level of reflexivity, such encounters might engender learning and empathy toward one another's beliefs and practices. In the case of AFM, there was little scope for reflexivity that might draw perspectives beyond religious doctrine. However, the same held partly true for me. As a white Canadian male, my own attempts to understand AFM and Pala'wan desires for reform often fell short when confronted by the missionaries' abhorrence of Indigenous customary practices and some Pala'wan desires to "leave them behind." Keeping an open mind about the mission's disciplinary practices was not easy.

9. The AFM filmed the videos in Kamantian and described in detail their church practices, medical facilities, recruitment processes, and thoughts on Pala'wan customs, and, importantly, what they had achieved and where else they planned to convert other "lost souls."

1. BIOPOLITICS, MATERIALITIES, AND THE POLITICS OF DIFFERENCE

1. Cepek (2011) critically questions how Agrawal's (1999) "environmentality" and Goldman's (2001, 501) "eco-governmentality" assume that local subjects' self-disciplining toward eco-rational behavior is inevitable and complete. Cepek argues that such work "underestimates the degree to which people are capable of forging a critical, self-aware and culturally framed perspective on collaborative projects for socioecological transformation" (501). In his study, Cofán participants maintained a "critical consciousness" of the activities and objectives of conservation and viewed their participation in relation to political aims and aspirations rather than any project goal (502). In Rubis and Theriault (2019), the Pala'wan also articulate a critical consciousness toward environmental rule enacted through "protocols of concealment." In a separate paper, Theriault (2017) similarly argues that Pala'wan ontological worlds, beliefs, and practices make aligning with the Mount Mantalingahan Protected Landscapes (so far relatively weak) governance regime difficult (Theriault 2017, 2019).

2. Biopolitics on Palawan is always sociospatially variegated in intensity and influence. For example, I found that male Tagbanua elders in interior spaces tended to pay less attention to rangers and NGO prohibitions against clearing old forest for swidden, often using their children's labor to clear and cultivate older forest areas for their "fat soil" (*mataba ang lupa*) and high rice yields. Like the elderly in Marenshewan, they expressed little concern for the prevailing antiswidden discourse on the island. In contrast, middle-aged Tagbanua men living closer to the mixed forests and paddy fields of the CADC buffer zone knew that the practice was illegal and sanctioned. They cultivated swidden less frequently and in young secondary growth. Many of their leaders also monitored swiddens as active "brokers" and "translators" (Dressler 2013) for NGOs and a national park. Over the years, the combined pressures of governance, livelihood changes, and declines in soil fertility have led many farmers to stop clearing and burning. Instead, most choose to underbrush or plow their fields in response to a combination of fines, monitoring, land shortages, new crop markets, and diminishing forest areas available for clearing. Antiswidden discourses influence people and places with varying degrees of intensity, depending on their involvement in governance, the level and type of disciplining and incentives, and, crucially, how their lives and livelihoods are affected by uneven agrarian political economies.

3. Bhattacharya et al. (2017, 6–7) consider social reproduction to be open, unfinished, and processual, involving "activities and attitudes, behaviours and emotions, and responsibilities and relations directly involved in maintaining life, on a daily basis and intergenerationally. It involves various kinds of socially necessary work—mental, physical, and emotional—aimed at providing the historically and socially, as well as biologically, defined means for maintaining and reproducing population. Among other things, social reproduction includes how food, clothing and shelter are made available for immediate consumption, how the maintenance and socialization of children is accomplished, how care of the elderly and infirm is provided, and how sexuality is socially constructed."

4. Netting (1993, 100–101) offers a standard definition of the rural household as one that "mobilizes and allocates the labor and manages the resources of the smallholding: the household is the key productive unit. Though household members may also carry on individual agricultural production or have nonfarm occupations, they generally contribute to the farm enterprise in material ways and derive a part of their consumption from pooled household subsistence production and income."

5. Standard societal (and disciplinary) interpretations of peasant smallholder "households" and "farms" include belonging to a corporate nuclear family unit, holding productive assets, engaging in sedentary agriculture or productive enterprises, and acting as landed entrepreneurs (Netting 1993, Wilk 1989, 1991; Ellis 1993). Such enduring interpretations inform the blueprint of nonstate interventions to reform and modernize uplander lives and livelihoods in line with normative household attributes, including sedentarism, economic rationalism, and hygienic practices.

6. The analysis of class and differentiation among the Pala'wan themselves must also be approached with caution. Social reproduction unfolds along "horizontal" social relations of sharing and reciprocity that are less geared toward accumulation and commodification processes (land, labor, and capital), although this is changing.

7. Such dynamics of social reproduction include individual, familial, and kin relations, livelihood multiplicity (e.g., mixed subsistence, cash cropping in swidden, hunting and collecting NTFPs, intermittent daily wage labor in lowland paddy fields, and selling tingkep baskets), sociocultural practices, political maneuvering, and everyday life in the hinterland, coasts, and urban areas.

8. At the interstices of forest and agrarian livelihoods, most Pala'wan social relations and livelihoods remain farm and forest reliant, either out of preference or because they are ostracized from controlling, or investing in, formal landholdings, commercial agriculture (e.g., titles, assets, and capital), or other lowland economic opportunities. Moreover, many Pala'wan families avoid or fail to enroll in the provincial public schools or deregister early, whereas others avoid lowland institutions (even hospitals) altogether. Such avoidance is often deliberate. Many Pala'wan endure discrimination from Christian lowlanders, including evangelical Christians.

9. Missionary control of "sacred materialities" was, of course, far from immediate. Hefner (1993, 104) notes, for example, how a small number of Javanese and Euro-Javanese established Christian villages in the lower Brantas River Basin on Java in 1830, which allowed for the comingling of Muslims, Christians, and Javanese custom in the teaching of Christian beliefs. While village residents had to conform to a minimum of Christian strictures, these strictures coexisted with Javanese "mystical terms" and material culture (e.g., Gamelan music). It was only when Dutch missionaries heard about these "syncretic" Christian villages that village residents were brought under the authority of the Dutch Reformed mission and its harsh measures (e.g., prohibiting Indigenous dress, arts, names, and language).

10. As Jones (2002, 399) describes, Episcopal missionaries depicted the Indigenous "Igorot" in the mountains of northern Luzon between 1903 and 1916 as culturally

backward peoples to justify their efforts to educate and curb customary head hunting practices associated with appeasing malevolent spirits. Under the Spanish, the Episcopal missionaries frequently commented on how the Igorot peoples held steadfast to customary social relations and material practices over decades of religious and military colonialization (McCoy 1982).

11. Such agentive capacity is neither intrinsic nor inherent to the materiality of the tingkep or other upland items. Rather, the basket's capacity to influence emerges through its social and material form and usage across different social spheres in upland and lowland settings.

12. Indigeneity is often easier to comprehend in the context of perceived distinctions between "First Peoples" and European colonizers in places like North America. Some national governments in the region (e.g., Vietnam) completely reject the term, suggesting that all their citizens have "first-order" connections with people and places (Baird 2016). Other scholars suggest that autochthony is less useful in understanding indigeneity, as more peoples are migrating, displaced, and relocated onto new lands for biopolitical ends, potentially nullifying any first-order conceptualizations. Consequently, many suggest that Indigeneity, "being Indigenous," is best understood in terms of specific criteria or conditions that mark a people as "Indigenous" (e.g., customary practices and territory) and relational factors between "Indigenous" peoples and other actors (e.g., marginalization) in changing political-economic contexts (Merlan 2009, 305; Baird 2016). I adopt this broader line of reasoning in engaging with Indigeneity on Palawan Island and the Philippines.

13. As part of the West's postwar "liberal turn," global multilateral institutions (e.g., United Nations, World Bank, United Nations International Labor Organization), transnational NGOs (e.g., International Union for Conservation of Nature, World Wildlife Fund), and nation-states pivoted in the late 1970s toward greater political recognition of Indigenous peoples' rights to land, resources, and self-determination (Povinelli 2002; Niezen 2003; Chapin 2004; Igoe 2005; Merlan 2009).

14. Igoe (2005, 3) sees the circulation of Indigenous rights discourse and participation in conservation as a form of "global indigenism" (see also Merlan 2009)—one that has saturated the most remote of hinterland areas.

15. This shifting eco-Indigenous middle ground "was founded on the assertion that native peoples' views of nature and ways of using natural resources are consistent with Western conservationist principles" (Conklin and Graham 1995, 696). Indigenous leadership and communities will sustain such partnerships by negotiating and appropriating the politics, ideas, needs, and concerns of various actors in such alliances. As adept brokers, they may actively leverage "traditional" knowledge and practices to support community needs while bureaucratically enabling the NGO's project aims and objectives (Hirtz 2003; Fletcher et al. 2019).

2. UPLAND LIVING AND TINGKEP WORLDS

1. It is beyond the scope of this book and my knowledge of Pala'wan society to offer an in-depth ethnographic account of Pala'wan cosmology, faith, and beliefs across human and nonhuman realms. I therefore supplement my own ethnography with the most comprehensive and detailed accounts of Pala'wan worlds from the meticulous work of Charles Macdonald (1992, 1997, 2007, 2011); Nicole Revel (1990, 1998, 2017); Dario Novellino (2001); Will Smith (2015, 2021); and Noah Theriault (2017).

2. My experience with Tagbanua *babalyan* suggests that they very much continue to uphold degrees of authority, leadership, and reverence.

3. In highland Palawan, the names Empuq Banar and Diyos (Jesus) are often used interchangeably in discussions and interviews.

4. Broader human-environment relations (e.g., climate variability, landslides, or flooding) are the realm of Empuq Banar, whose appeasement with sacrificial ritual mediates and recalibrates ruptured relations (see Smith 2015, 2018, 2021).

5. Often synonymous with diwata, *taw kewasa* are "powerful beings" or "spirits or divinities living in the sky (*dut langew*)" or higher in the mountains/ mountain peaks (Macdonald 2007, 102). Closer to Empu's abode, they are described as benevolent and helpful.

6. *Ungsud* is a broadly used term that refers to a ritual offering and/or sacrifices of various sizes and characteristics. It generally involves important sociomaterial offerings with soul (kurudwa), such as rice or a white chicken, to appease diwata or taw't talun and to mediate socioecological uncertainty in the human-nonhuman realm. Macdonald (2007) notes that the term is also used in reference to bride-price.

7. The Pala'wan farmer Mandoso Burak, who lives in the Rizal interior near the Kamantian uplands, recounts how incest between a sibling couple caused Empu to sacrifice their children, whose body parts are scattered across a field to yield rice as the basis of their lifeworld. In this way, uma rice production becomes central to nourishing the lives of Pala'wan, more so than root crops, forest edibles, and fruits, which, as Burak notes, predate the arrival of rice. He explains:

"It was like this, there were twins who got married. Back then, a twin sibling couple got married and had twins. But Empu said, 'You cannot be married, it is prohibited, so your children will bear the burden of punishment, and be killed by [the female figure] Maraga [who Empu had directed in a dream].' Once Empu arrived, he then told them that their children's death will be the source of our livelihood. Then Maraga chopped the twins into small pieces and Empu scattered their body parts across the uma. From the scattered body parts, emerged the parey [rice], and that, it was stated, was the origin of the parey. It was only banana and cassava that served as the source of our livelihood until Empu said, 'I am offering you another livelihood. I want you to farm uma as your source of food.' That is why our ancestors started planting rice, for us to be full, so we won't go hungry. So parey became our main source of life. That is why, if you notice, every time when you harvest or process rice, the rice speaks in a soft voice, 'Ouch (*adey/aray*),'" because of the pain inflicted by Maraga upon the twins. Even when you pound rice, it is heavy work, you experience this pain as the pounding continues. That is why, we should cultivate parey with great care—it is the source of life from Empu."

8. Often mediated by beljan, Ginawa can be a source of health or illness (never both) from animate (plants and animals) and inanimate objects (*mutja*), which can repel and or cure illness (Dressler, fieldnotes).

9. Iskander (2016, 5) notes that, concerning the "afterlife," her informants describe Empu judging and guiding souls to *langit* (a place of light) when considered "good" and *narka* (a place of fire) when considered "bad."

10. Empu also imbues other human and nonhuman entities with *ginawa,* breath of life, with agentive properties (Macdonald 2007, 124). Depending on the object, ginawa animates both human and nonhuman entities, including the varied mutja (talismans), such as stones and floral and faunal parts, placed inside tingkep and the forest materials used to make tingkep and so animates the basket with degrees of agency and power (also Iskander 2021). However, if certain divine objects are placed or captured in the tingkep, such as uma rice, then Kurudwa may be at play, bringing consciousness and volition to material objects, invoking diwata and other entities. Kurudwa and ginawa thus reflect the diversity of life, spirit, and materiality that saturates the ontological realm of forests from which the tingkep and other customary objects are sourced and in which they are embedded.

11. In this story and other narrations, the beljan uses a sprig of ruruku to "foster clairvoyance" to see through the "invisible and to forge a path." Wafting the smoke from parina

resin or wood "makes the world transparent and allows the shaman to move through the visible and the invisible during his Voyage" (Revel 1998, 2).

12. I have witnessed many multiday Tagbanua healing ceremonies (*pagdiwata*) involving *Babalyan* entering into repeat trances to negotiate with *panya'en* (the Tagbanua equivalent of lenggam on how to cure an illness brought about by ruptured or stressed human-environment relations—for example, cutting so deep into an almaciga tree (*Agathis philippnesis*) that the exuding resin resembles human blood, which causes *panya'en* to unleash their demonic pets (*damdam*) to inflict illness and death on the harvesters (see Fox 1954).

13. In other cases, farmers may use a machete to etch an "X" into a tree to signify the corner or lateral boundaries of a field. Although the practice is less common in highland areas, midland and lowland farmers often delineate and claim swidden fields to defend increasingly scarce holdings from migrant homesteaders and other private actors attempting to expropriate Pala'wan lands and resources.

14. Revel, Xhauflair, and Colili (2017) refer to the season of felling trees as *panambang* and the custom of mutual help and assistance as *Adat et tabang*.

15. If Binawagan is not seen alongside Maroposo, farmers will avoid planting, as they anticipate the untimely arrival of heavy rains. Both constellations leaving, or no longer being visible in the night sky, heralds the arrival of *pinaburukan-*: a period of heavy, soil-soaking rains. During this time, a farmer (or beljan) may ritually sacrifice a white chicken, along with honey, tobacco, and rice wine (*tinapey*) to appease Empu and restore sun and heat. This may bring or fall in line with the time of balikwat (*balik*, to return), which marks the return of Maroporo above the horizon and a short period of lighter, intermittent rains (the arrival of a small or young barat) followed by heat during which uma planting can continue before the barat rains arrive (Revel 1990).

16. Typically, a woman's domain includes smaller vegetable gardens at the edges of swidden fields or houses. Near the farmers' home, or temporary swidden (guard) house, one finds an assortment of vegetables, including eggplant (talong, *Solanum melongena*), okra (okra, *Hibiscus esculentus*) and other leguminous plants used for flavoring, including chili pepper (Zili, *Capsicum frutescens*), ginger (luya, *Zingiber officinale*), and pigeon peas (kadios, *Cajanus cajan*). Other tree crops—such as various banana (saging), papaya (kapayas, *Carica arabica L.*), jackfruit (lanka, *Artocarpus integer M*), mango (manga, *Mangifera indica L.*), and Philippine lime (kalamansi, *Citrus microcarpa L.*)—may also find their way into swidden fallows. With the ability to cut, crop, and move stalks of taro, usually from the wetter soils of a swidden, banana and cassava (usually from drier soils) farmers can transplant shoots and stalks in village clusters. In other cases, a farmer can also draw on maturing tree and root crops in older swidden fallows, often considered a storehouse in times of scarcity.

17. Like other Indigenous highlanders, the Pala'wan in Kamantian and Marenshewan now use far fewer varieties of heirloom rice and practice swidden ceremonies much less frequently. Instead, many invest in individual or family-based ceremonies before planting their umas—for example, placing "first rice" in the middle of a field under lemongrass (*tanglad*) or in the corners or edges, accompanied by a brief prayer in the uma to Empuet Parey to ensure the vigorous growth of rice.

18. Depending on the varieties grown in the uma, the swidden rice harvest may be staggered throughout the swidden season. Shorter maturing varieties are harvested three months after planting, and longer maturing varieties take four to five months; they may be planted successively or in a different section of the field (see Macdonald 2007; Novellino 2011; Smith 2015).

19. The cumulative pressures on access to and use of ancestral forest lands, swidden, and declining rice yields vary from place to place. However, they usually include

a combination of factors, such as in-migration and population increases, conservation interventions and zoning, access to lands and reduced fallow periods, declining soil fertility, the investment of less labor in field maintenance, and the allocation of more time to other livelihood activities in the lowlands.

20. Unlike their lowland counterparts, few households in Kamantian and Marenshewan have the land, income, and capital—for example, draft animals (*carabao*) and pesticide—required to undertake irrigated paddy rice cultivation (*basakan*).

Those Pala'wan who have succeeded in claiming relatively flat to undulating lands, or managed a terrace system, often shift to more bountiful (irrigated or rainfed) wet rice production that, in some instances, can yield two to three croppings each year, with yields significantly greater than swidden rice cultivation. As Macdonald (2007, 50–51) notes, and I have observed, Indigenous lowlanders with regular, longer-term access to secure land tenure (based on tax declarations or private deeds) tend to leverage loans and capital, different types of cash crop production (e.g., regular copra production and sales), and/ or consistent labor opportunities that enable them to invest in, and overcome the establishment costs, of more capital intensive paddy rice cultivation (see also Dressler 2009). However, shifts to more intensified commercial agriculture seldom happen alone—most Pala'wan rely on relatives or migrant farmers to build up their networks, knowledge, and productive capacity.

21. In contrast to the Pala'wan in the mid- to lowland areas, those in Kamantian and Marenshewan do not engage heavily in copra production. In most cases, although it is a low capital activity, sufficient stands of coconut trees are required to make the splitting and drying of the pulp and selling of copra worthwhile (see Macdonald 2007). The cooler, wetter forested upland areas are far less suitable for copra plantations, with only a few coconut trees being used for the milk, bark, hair of nut, and husk for fuel, swidden firing, and smoking out of bees for honey harvesting.

22. Rattan is described as a woody, climbing palm. Rattan's woody mid-stem and its derivatives are used to construct cane and wicker goods such as baskets and furniture (De Beer and McDermott 1989). The two major genera in Palawan are *Calamus* and *Daemonorops* (PCARRD 1991). Most species harvested commercially in Palawan are from *Calamus*. Dozens of rattan varieties are harvested in the uplands and lowland of southern Palawan including but, not limited to, larger diameter species such as *uway* (*Calamus maximus Blanco*), *bugtong* (*Calamus subinermis*) and *seka* (*Calamus microsphaerion Becc*) and smaller diameter endemic species *pin pin* (*Daemonorops margaritae* (Hance) var *palawanica becc*) and others used in basket weaving (see below for the species used for tingkep, and PCARRD [1991] for the sixty-six known rattan species in the Philippines). In most cases, Indigenous highlanders use the names *yantok* or *uway* as generic labels for the larger, better-priced (and increasing scare) rattan sold in the lowlands.

23. Pala'wan tap almaciga resin from the bark of the highland forest tree, *Agathis philippinensis* (De Beer and McDermott 1989). In trade parlance, it is known as "manila copal" and used as a varnish or paint for furniture after being processed. Groups of five or so Pala'wan men travel to inherited ancestral trees; they make small three- to four-inch incisions at a depth of about 2 cm, at 70–160 cm basal diameter height. After waiting at least two months, they chip off the accumulated resin from the incision. The group of harvesters can collect 115 to 130 kg of resin from a dozen bagtik trees, with each carrying over 20 kg of resin down in their *kiba* (homemade rattan carrying basket) to weighing and buying stations in the lowlands (Bibal 2021). *Tipak*—a whiter, translucent almaciga resin—is the standard grade resin harvested after three to four months of initial tapping. *Binunga* is the most prized amber-colored resin, harvested after two to three years of tapping (85). While the volume of the harvest depends on the health and age of trees, the extent of tapping, and weather in the leadup to and during the harvest, the final income is affected

by debt from food loans and how much the buyer charges for "bodega" maintenance and organizational fees (81). The final "forest gate" payment amount is, therefore, about 20 to 30 percent less than market rates, making the bagtik harvest as precarious as swidden, but perhaps more important, since it generates a much needed cash income.

24. *Mutja* are often used during the hunt or harvest. The Tagbanua also use recently found or inherited mutja (known as *bangat* or *banget*) during pig hunting, honey harvests, trapping, and fishing. In 2001, now deceased Babalyan Librito Matuar in Napsaan noted, "You need mutja to find the beehives. You need to prepare your mutja if you want to find the honey. We call on different spirits when we do uma or gather honey. You will also need something that will tame the animals. You could get the parina from the sap of the trees. Sometimes, we would put the parina inside the fish and use this to tame the animals. . . . If you have mutja, you would be able to kill at least two pigs in a flock, either with your gun or with a spear. If you use a spear, you will surely kill two pigs." From Cabayugan, elder Fabio Franco supported this by saying: "Before hunting the pigs, we use herbs from the forest and rub them vigorously on the snout of the dog or are fed to the dogs to make them brave. When the dogs are brave, we then make them smell fragrant wood so they can smell the wild boar, and it will be easier for the dogs to find them. This is what is called *banget*, which makes dogs brave. We also use *pangter* for the pig to make it weak." The mutja imbues the owner and user with the powers required to overcome uncertainty across a range of social relations and livelihood practices in ways that cut into the everyday and spiritual realm.

3. SPANISH COLONISTS, FORESTS, AND GODS

1. Earlier legal provisions under the Spanish Law of the Indies and mounting discontent among Indigenous peoples in reducciones enabled partial (and rather fleeting) recognition of Indigenous conceptions of ownership based on occupancy, clearing, and cultivation. The law initially governed Spain's possessions in the Philippines and, from 1523 onward, clarified that any land rights granted to "loyal Spanish subjects was not to impair the rights and interests of the natives to their holdings" (Lynch 1982, 274). This would open the door for the contemporary recognition of ancestral domain claims and titling (Dressler 2009). Despite subsequent decrees supporting the Law of the Indies, the Crown and state continued to use the Regalian Doctrine to relinquish upland people's right to access and use public forests. These decrees stated that long and continuous possession qualified Indigenous peoples for a title, yet supporting legal provisions were not always considered by the colonial courts. Uplanders' methods of articulating their hold over land—based on a history of occupation, use, and beliefs—conflicted with the colonial state's interest in upland forest landscapes for timber and territory.

2. Warren (2007) notes that food shortages were not uncommon in Jolo due to the island's limited agricultural potential (e.g., its erratic rains), which limited irrigated or rainfed paddy rice systems. Additionally, many Tausug chose not to participate in rice production because its seasonality conflicted with maritime "trade and procurement activities" (95). Consequently, "the Tausug looked to their neighbours to solve their problems of chronic food shortage . . . relying on [tribute] rice imported from the Cotabato basin, stapling points on southern Palawan, Basilan" and elsewhere (97).

3. Frustrated, the Spanish colonizers increasingly used the pejorative term *Moro*—drawn from the Spaniards' fifteenth-century exonym for Muslims of North Africa, the Iberian Peninsula, and parts of the Mediterranean—to refer to the local Muslim populations who resisted subjugation with military force (Mckenna 1996; Paredes 2013). In Mindanao, Filipino Muslims have reappropriated the term in their social and political struggles for an independent homeland. In 2019, this struggle culminated into the

Bangsamoro Autonomous Region in Muslim Mindanao (BARMM) following negotiations between the Philippine state and the MILF (Hall and Deinla 2019; see Castillo 2014 for the historical developments of the BARMM).

4. AMERICAN FORESTERS, NONSTATE RULE, AND THE TRIBAL "OTHER"

1. In this book, I use the terms "American" and "America" to refer to the United States of America and use "United States" ("US") in specific reference to administration and bureaucratic practices to ensure greater specificity. For the sake of consistency, I avoid the use of "USA."

2. It was only after some time that American colonial foresters came to appreciate that, apart from the drier pine forests in Benget in northern Luzon, the moister forest conditions in the Philippines made it difficult for fire to spread from swidden plots. As timber concessions and deforestation rates grew, reforestation efforts failed to sustain timber supplies, leading the American colonists (and, later, the Philippine Commonwealth and Republic) to designate numerous upland mountain areas as forest reserves and national parks (Hurst 1992). In this context, American colonial and Philippine forestry officials sustained what Bagadion (2000, 119) called a three-pronged "plant-punish-police" approach to forest reserves and national park management. Nonstate actors would replicate the approach decades later.

3. The Bureau's Forest Act (1904, 15), Section 25, stated that "the cutting, clearing, or destroying of the public forests or forest reserves, or any part thereof, for the purpose of making caingins [swidden], without lawful authority, is hereby prohibited, and whoever, in violation of this provision, shall cut, clear, or destroy the same, for such purpose, or shall wilfully or negligently set fire thereto, shall, upon conviction by a court of competent jurisdiction, be punished by a fine not exceeding a sum equivalent to twice the regular Government charge upon the timber so cut, cleared, or destroyed, and, in addition, thereto, by imprisonment not exceeding thirty days, in the discretion of the court."

4. Bureau of Forestry, Forest Manual, Section 25, notes: "(a) The clearing by fire of grass and brush land and of land in the public forest containing but little timber of inferior species, for the purpose of making 'caingins' will be permitted in the following manner only: Application will be made in writing to the local forest officer, or in his absence to the nearest president of a municipality or settlement, for permission to make a 'caingin'" (45).

5. For the exchange relations of nondomestic protein (and other NTFPs) between upland and lowland farmers, see Peterson 1978 for Agta-Palanan and Dressler 2009 for Tagbanua-Visaya.

6. The Pala'wan and other Indigenous peoples labored in uneven patron-client relations for rice to repay accumulated debts in foodstuff and to receive cash since greater quantities of surplus rice were available in the lowlands. This arrangement exists to this day (Smith 2015). The Muslim Tausug's long-term engagement in fishing and copra production along the coastal plain was supplemented with—and in some cases replaced by—paddy rice production (Smith 2015; Macdonald 2007). Today, paddy fields often flank large tracks of mature coconut stands planted by Pala'wan, Muslim, and Christian migrant lowlanders in the 1950s–1960s for copra production. Pala'wan also have their own coconut groves further upland, which are used in diverse ways (e.g., palm fronds used in house walling and roof thatching; timber from the trunk for flooring and walling; coconut oil, flesh and milk for family consumption and sale, or pig feed; coconut shell rendered into charcoal; and coconut husk used for fire starter or to smoke out bees; among many other uses).

7. According to Paredes (2013, 25), the Cebuano term "Lumad" is a modern political term that refers to the Indigenous peoples of Mindanao who did not convert to Islam prior

to, during, or after the Spanish colonial period." She adds that it "literally means "born from the earth," and, despite its etymology, it is widely used for collective self-reference across eighteen culturally related—but nonetheless distinct—ethnolinguistic groups. In other words, it identifies an ethnic category (not a single group), which Leonard Andaya defines in his history of Malayu ethnicity" (25).

8. Religiously or ceremonially "clean" animals include fish with scales and fins, chickens, duck, geese, and four-legged animals that chew the cud and have split hooves, such as water buffalo, sheep, and goats (Banta et al. 2018).

9. These SDA missionaries also sought to establish a printing press upon their return and raise funds for a meeting place to accommodate at least three hundred people for Evangelical sermons near Manila (Fernandez 1990).

10. The AFM's biopolitical reach was geographically circumscribed in a so-called 10/40 window—that is, an unreached, pagan region "which stretches from northeastern Africa across southern Asia between 10 and 40 degrees north latitude" (AFM 2021).

11. In 1974, the Forest Occupancy Management Program (FOMP) further enabled state forester mandates to settle the kainginero and stabilize their swidden farming systems. In other cases, the FOMP legitimized the resettlement and relocation of swiddeners by offering them land lease opportunities and permanent agriculture. As specified in PD 705, this involved rendering swidden cultivators sedentary "fixed plot" holders by granting them a short-term "occupancy permit" to contain fields (of up to seven hectares) within specified boundaries. To prevent swidden from encroaching on forest lands, "qualified forest occupants" could remain on unclassified public lands on the condition that fields be reforested and intercropped on a permanent basis (Population Center Foundation 1980, 45). Those who illegally felled trees were supposedly given amnesty from prosecution.

12. Most program initiatives eventually fell under the 1995 Community-Based Forestry Management Agreements umbrella initiative. They only offered partial recognition of upland peoples' usufruct holdings on twenty-five-year renewable leases, provided they curb swidden farming through agroforestry and related practices (Dressler and Pulhin 2010). Such "people-oriented" programs were also often seen as counterinsurgency measures to "maintain political stability and order in the countryside" (Pulhin, 2000, 2) as the Communist Party of the Philippines and New Peoples' Army mounted a bloody insurgency against the Marcos administration.

13. Both religious and "secular" nonstate actors—the latter of whom also had religious inclinations—facilitated decentralized political reforms by supporting the various environmental and human rights movements emerging against Ferdinand Marcos's oppressive regime. In particular, the Catholic Church, witnessing the worsening conditions of the poor and intensifying extrajudicial killings, reemerged as the "Church of the Poor" and sanctioned the toppling of the Marcos government (Holden 2012).

14. According to IPRA: Chap. 2, Sec. 3 (h): "A group of people *or homogenous societies identified by Self ascription and ascription by others, who have continuously lived as organized community on communally bounded and defined territory*, and who have, under claims of ownership since time immemorial, occupied, possessed and utilized such territories, sharing common bonds of language, customs, traditions and other distinctive cultural traits, or who have, through resistance to political, social and cultural inroads of colonization, non-Indigenous religions and cultures, became historically differentiated from the majority of Filipinos. And that: ICCs/IPs shall likewise include peoples who are regarded as indigenous on account of their descent from the populations which inhabited the country, at the time of conquest or colonization, or at the time of inroads of non-Indigenous religions and cultures, or the establishment of present state boundaries, *who retain some or all of their own social, economic, cultural and political institutions*, but who

may have been displaced from their traditional domains or who may have resettled outside their ancestral domains."

15. Given that CADTs and similar tenurial systems seldom hold the same legal recognition as state or private land control, they often act more as cultural beacons for nonstate actors implementing reform projects (e.g., conservation, missionizing, or other extractive purposes) than a legal instrument that buffers political and economic encroachment (Dressler 2009).

16. As the Philippines was declared a middle-income country under then-president Gloria Macapagal Arroyo (and her intense campaign of deregulation), both NGOs and the state experienced a decline in donor aid "as bilateral priorities began to shift and foreign aid budgets fell" (Parks 2008, 44). Grants from bilaterals decreased significantly from $296.5 million to a mere $165.9 million within the same period. Domestic NGOs suffered significantly, with 50 percent to 95 percent of their annual budgets coming from grants (675). Even as many NGOs increasingly relied on staff to volunteer labor, the state's own budgets were running dry and the government leaned on NGOs to work in remoter parts of the country, including southern Palawan.

5. OF FORESTS AND GODS

1. As noted, most sought out new lands and opportunities to settle, start families, and invite relatives to join them, comingling with long-term Tausug and other Muslim lowland settlers who had already cleared lowland forests for copra and paddy rice production (Brown 1990).

2. Most migrants started with swidden and transitioned into paddy rice (basakan) and copra plantations in the lowland or midland areas. They have only recently expanded upland into Pala'wan territories. Pala'wan and other Indigenous uplanders increasingly rely on irregular wage labor on migrant-owned paddy fields and copra plots in the lowlands, as they draw on their own lands, forest, and water for varied subsistence and commercial uses. In lowland areas, most Pala'wan have access to loans, capital, titles, or land suitable for commercial production. In terraced midland areas, most paddy fields are small (0.5–2.0 ha), with little tenurial security. While surplus is higher than swidden, yields are only just enough to feed a family for the year. In this sense, despite ethnolinguistic and spatial markers of difference, Pala'wan social relations and livelihoods have long intersected with and been influenced by Muslim and Christian lowlanders. In earlier times, this intersection occurred within the context of relative Pala'wan autonomy and comparative upland advantage.

3. NATRIPAL (Nagkakaisang mga Tribu ng Palawan) is technically a PO but functions as an NGO. It represents a province-wide federation of fifty-seven or more local associations in thirteen or more of Palawan's twenty-four municipalities (Pinto 1999).

4. In some cases, it was even reflexive. Indeed, the Apostolic Vicariate of Puerto Princesa and CSSC reflected on the impact of the language they used to describe Indigenous uplanders. As the vcariate's own webpage notes: "The name Indigenous Peoples Apostolate was introduced as replacement to Tribal Filipino Apostolate in consonance with the declaration of the United Nation on its preference to use Indigenous peoples rather than Tribal Filipino to discourage the marginalization that the term Tribal Filipino implies." "Apostolic Vicariate of Puerto Princesa," Union of Catholic Asian News, accessed February 1, 2021, https://www.ucanews.com/directory/dioceses/philippines-puerto-princesa/470.

5. National and international social and environmental movements achieved a nationwide logging ban in 1993 under Fidel Ramos's administration (1992–98). One campaign

was particularly instrumental in curbing deforestation. The Haribon Foundation—an influential environmental NGO from Manila—used a million-signature petition to successfully lobby the preceding Aquino government to declare a twenty-five-year logging ban in Palawan (Vitug 1993). A protracted and acrimonious debate between local environmentalists and businessman Jose Alvarez and his companies quickly surfaced in national newspapers. Each side expounded on and upheld their respective positions: ecosystem integrity and land rights versus "sustained" employment, respectively.

6. Localized water supply systems that could feed well water to mid-elevation hamlets to encourage uplanders to gradually sedentarize and move to the lowlands. The aim was to encourage uplanders to relocate from upper elevations and away from environmentally critical ECAN zones (see Dressler 2013). According to Hobbes (2000), however, many Pala'wan further upland resented the PTFFP for its strict antiswidden rules and for influencing participants to adopt Christian lowlander ways and means, such as modern clothing. Pala'wan living further upland resented the fact that they could no longer clear forest without the potential of fines, and others believed the PTFPP would seize their lands and force them to settle in the lowlands, leading to a loss of culture and assimilation (67). Such critical reflections by Pala'wan suggest that the PTFPP's biopolitical imprint was not always welcome.

7. "Mount Mantalingahan. Our Shared Goal: No More Forest Loss," CI Philippines, accessed July 24, 2024, https://www.conservation.org/philippines/projects/mount-man talingahan-protected-landscape.

8. "Mount Mantalingahan. Our Shared Goal: No More Forest Loss."

9. "Mount Mantalingahan. Our Shared Goal: No More Forest Loss."

10. "Mount Mantalingahan. Our Shared Goal: No More Forest Loss."

11. CI interview 2001, Director of Conservation International, Puerto Princesa City, Palawan Island, the Philippines.

12. The Protect Wildlife Project currently aims to reinsert state governance and policy—particularly the PCSDs and DENR—into the remote uplands. This involves a complex network of NGOs, corporations, local government units (Christian and Muslim), farmers, integrated conservation and development project on-site managers, and wildlife-environmental law enforcers. It also includes lowland and upland Pala'wan community members who, through locally organized social and virtual networks, will serve as conservation communicators (DAI 2021). Unlike earlier biopolitical interventions, the project uses the digital Biodiversity Resources Access Information Network (BRAIN) system to "innovate enforcement coordination and management and to promote efficiency in the agency's [PCSDs] regulatory processes" (DAI 2021, 173). A mobile and web-based application, Wildlife Insights (from https://www.wildlifeinsights.org/home), captures, communicates, identifies, and categorizes images of endangered species and connects the so-called rapid wildlife enforcement units, wildlife tracking monitoring units, and members of the Palawan Environmental Enforcement Network (PALAWEEN). The WildALERT mobile application allows citizens to report (to DENR and PCSDs) illegal wildlife poaching and trafficking, while the web-based RESPONSE system "share[s] real-time information with partners; and plot[s] enforcement routes, target[s] locations and entry and exit points of illegally harvested flora and fauna" (DAI et al. 2021, 173). Overall, this biomonitoring, security monitoring, and surveillance system seems to be effective, but the role of Indigenous Pala'wan, Tagbanua, and Batak remains uncertain.

13. J. Mariñas, "The Palawan Adventist Mission," Encyclopedia Adventist, accessed May 23, 2024, https://encyclopedia.adventist.org/assets/pdf/article-bart.pdf; "New Opportunities in Palawan. Adventist Frontier Missions, Reaching the Unreached Posted," Adventist Frontier Missions, accessed July 24, 2024, https://afmonline.org/post/new-opportunities-in-palawan.

14. "Why We Go," Adventist Frontier Missions, accessed September 15, 2018, www.afmonline.org accessed September 15, 2018.

15. For example, after lighting cheap cigars to celebrate my birthday near the pastor's house, I was told repeatedly that smoking and drinking were out of the question.

16. Every year, a new crop of young AFMs, typically Americans, come to work in the Kamantian area. Their blogs detail "watch[ing] for the transformation of God . . . working in their lives and in their homes . . . [to see] another witchdoctor being converted." "Eyewitness," Adventist Frontier Missions, accessed September 12, 2018, https://afmonline.org/post/eyewitness.

6. FOR THE SAKE OF FORESTS

1. Rather than contracts being forsaken and replaced when one side fails to fulfill its side of the agreement, the mutual expectations and failure to deliver linger in the uplanders' memories, making it more difficult for another contract to be drafted in the next instance (O'Leary Simpson and Pellegrini 2022).

2. In some instances, the fixer and trader collect many baskets and drop them off at shops without being paid, hoping that they will be remunerated the next time around.

3. Some traders and shop owners even suggested that, while the baskets were still sought after in PPC, they no longer readily bought them due to an oversupply of stock and lower rate of clearance. Apparently, the tingkep market had saturated, with fewer potential purchases in Kamantian.

4. As I show, other entrepreneurial Pala'wan also work as local fixers who collect tingkep for traders and take a small cut per basket.

5. Those weavers to whom I spoke suggested these values were higher than normal, with larger bulk orders raising their return on a one-off basis. While most households produced smaller baskets, the few who had produced (5 hhs) medium to (5 hhs) larger-sized baskets earned ₱1,380 and ₱28,960, respectively, in 2013.

6. Several years after my initial interview with Malwang, we had another conversation in Bataraza in 2019 about how his craft and tingkep trading business was going. Apparently, he had scaled up his business and was now involved in microfinance, offering upland Pala'wan weavers loans to weave baskets and make other crafts on advance orders. He noted, "We loan them money [to weave or otherwise] and they repay within six months, after which we can lend again. We borrow from lending companies like ASA, but not so much, like ₱20,000 only." He noted that repayment was based on baskets made and sold, with returns from sales going toward loan repayments. While he and others had a lot of success selling *walis tambo* (soft tiger grass brooms) and tingkep, there was no mention of how the indebtedness from loans to fellow Pala'wan in Mareshewan, Bono Bono, and Kamantian was impacting livelihoods in the uplands. While interest on such loans was relatively low, debt does not sit well with already meager household budgets in the highlands. It seems that nonstate endeavors to optimize Indigenous livelihoods have taken on an entirely new level of income financing and debt. Whether this would lead to less reliance on forests and more sustainable livelihoods is debatable. Many poor Indigenous households harvest forest resources more intensively to repay debt.

7. FOR THE SAKE OF GODS

1. The AFM missionaries bought the basket themselves and considered its production for tourism sales to be a potential means of social and economic betterment, not necessarily a form of cultural revitalization.

2. Apparently, the use of the helicopter was intermittent and subject to considerable maintenance. Recent reports are that the helicopter and crew crashed (and perished) during a mission off the coast of southern Palawan. See "Missionaries Lost at Sea," Adventist World, March 15, 2023, https://www.adventistworld.org/missionaries-lost-at-sea/.

3. The land of the dead is often characterized as a basin or valley, where the deceased may continue to live as normal. The souls of the deceased or the deceased may dwell in *kelebegang*, while those who sin go to *narka*, a "hell where they burn in a huge fire" (Macdonald 2007, 106).

4. Adventist religious restrictions on consuming wild boar (*biek talun*) have the potential to sever the social, material, and spiritual basis of sharing and generalized reciprocity across human and nonhuman worlds. The moral consensus and undertones of the need to share surplus meat and other items, particularly rice, remains strong in many upland communities. Macdonald (2007, 77) notes, for example, that after the hunt, the sharing of wild pig meat plays an important role in reinforcing and consolidating the "unity of local group and its kinship-based structure." Pigs were previously hunted by hunting parties with spears and dogs or trapped in a wooden spiked trap, *balatik*, or *bawag* (spring-loaded, bamboo spear trap) that are placed along trails. Pala'wan today often hunt boar with homemade guns and pig bombs, explosive devices mixed with gun powder, stones, shards of glass, and nails (if available) and baited with fruit or root crops from swidden fields. The bombs explode and the Pala'wan collect the dead, dismembered pigs. Portions of the pig are consumed by the lead hunters' family and then divided into smaller portions that are shared with kin and neighbors in other household clusters (2007). Pressing income needs can outweigh moral obligations to share, prompting the Pala'wan to sell meat to Christian migrants or wealthier Pala'wan in the lowlands. Overall, the redistribution of pig meat invests and affirms Pala'wan social relations, kinship ties, and local group structures. Wild boars figure prominently in the Pala'wan mythos and spirit world (see Smith 2020). Across the southern uplands, prominent Pala'wan mythology recounts complex relations between pigs, humans, and the nonhuman world of the Master of Pigs (Empu't biek), a spirit of the forest world (Macdonald 2007). In Pala'wan worlds, wild boars are considered "taw," human beings with kurudwa (a soul) and thus personhood. The interplay between these sociospiritual spheres reflects reciprocal predator/prey relations (or "generalised hunt," Revel, 1998) involving evil spirits (satjan or lenggam) preying on humans and human hunters preying on invisible humans (pig doubles; 121–22). Empu't biek relates how shifting between human and animal (pig) form unfolds in varied contexts, from everyday hunts, particular dreams, and the spirit world (see Theriault 2017). In the everyday realm, hunters may pursue a wild pig, which, when trapped and speared, emerges as an injured woman, who transforms herself back into a pig with a spear dangling from its back. In the chase, a pig may also transform into man and prevent a kill from unfolding. In my experience, the transformation of wild boar into a human form is manifested as evil forest spirits. In this sense, as Empu't biek offers up the pig to the hunters, Empu and these spirits must be respected or appeased before, during, and after the hunt. A lack of respect by not making offerings, damaging the pig's habitat, unrestrained killing of pigs or, in some areas, selling of pig meat for cash, leads to the rupture of reciprocity and ill fortune in the future (Revel 1998; Macdonald 2007; Theriault 2017).

5. Cockfighting is most often associated with lowland Christian migrants but has long been part of fraternal social activity among highlanders who have mingled with Christian and Muslim lowlanders.

6. In the lowlands, the SDA/AFM have begun buying properties near Brooke's Point with the intention of establishing a larger Adventist complex, complete with a lay training center, an evangelism center, and a storage unit. The complex will support their ministry in Kamantian and beyond, where, according to AFM pastors, there is no "Christian presence." AFM biopower thus explicitly seeks to bridge the lowlands and uplands in order to better extend inward and outward to "reach the unreachable."

References

ADB (Asian Development Bank). 1991. "Project Completion Report of the Palawan Integrated Area Development Project in the Republic of the Philippines." June 1991. Manila: ADB.

Adventist Frontier Missions (AFM). 2012. "Cholera." Adventist Frontier Missions Online. Accessed November 10, 2020. http://afmonline.org/post/cholera.

AFM. 2014a. "Field Update #8." AFM Field Reports. Accessed November 10, 2021. https://vimeo.com/107702764.

AFM. 2014b. "Field Update #7." AFM Field Reports. Accessed November 10, 2021. https://vimeo.com/107657694.

AFM. 2016. "Journey to Kebgen." Adventist Frontier Missions Online. Accessed November 10, 2021. https://afmonline.org/post/journey-to-kegben.

AFM. 2018. "He Fancies Himself a Witchdoctor." Adventist Frontier Missions Online. Accessed June 8, 2022. https://afmonline.org/post/he-fancies-himself-a-witchdoctor.

AFM. 2019. "An Introduction to the Remote Village of Kebgen." Adventist Frontier Missions Online. Accessed November 10, 2021. https://afmonline.org/post/an-introduction-to-the-romote-village-of-kebgen.

AFM. 2021a. "Kent & Leonda George: Career Missionaries since 1995, Serving Palaweno." Adventist Frontier Missions Online. Accessed October 3, 2021. https://afmonline.org/missionaries/detail/64.

AFM. 2021b. "About Us: Why We Go." | Adventist Frontier Missions Online. Accessed October 3, 2021. https://afmonline.org/about-us/why-we-go/.

AFM. 2021c. "About Us: AFM Church-Planting Model." Adventist Frontier Missions Online. Accessed October 3, 2021. https://afmonline.org/about-us/the-church-planting-cycle/.

AFM. 2021d. "Stumbling Block." Adventist Frontier Missions Online. Accessed June 1, 2022. https://afmonline.org/post/stumbling-block.

AFM. 2023. "Our First Missionaries." Adventist Frontier Missions Online. Accessed December 1, 2023, https://afmonline.org/about-us/our-history.

Ahearn, Laura. 2001. "Language and Agency." *Annual Review of Anthropology* 30: 109–37.

Agamben, Giorgio. 1995. *Homo Sacer: Sovereign Power and Bare Life*. Redwood City, CA: Stanford University Press.

Agrawal, Arun. 2020. *Environmentality: Technologies of Government and the Making of Subjects*. Durham, NC: Duke University Press.

Aguilar, Filomeno. 2005. "Tracing Origins: Ilustrado Nationalism and the Racial Science of Migration Waves." *Journal of Asian Studies*, no. 64: 605–37.

Anderson, Eric. 1976. "The Encomienda in Early Philippine Colonial History." *Asian Studies* 14, no. 1: 25–36.

Anderson, Warwick. 2006. *Colonial Pathologies. American Tropical Medicine, Race, and Hygiene in the Philippines*. Durham, NC: Duke University Press.

Anthias, Penelope. 2018. *Limits to Decolonization: Indigeneity, Territory, and Hydrocarbon Politics in the Bolivian Chaco*. Ithaca, NY: Cornell University Press.

Appadurai, Arjun. 1988. *The Social Life of Things: Commodities in Cultural Perspective.* Cambridge University Press.

Asiyanbi, Adenyi, and Kate Massarella. 2020. "Transformation Is What You Expect, Models Are What You Get: REDD+ and Models in Conservation and Development." *Journal of Political Ecology* 27, no. 1: 476–95.

Bagadion, Benjamin. 2000. "Social and Political Determinants of Successful Community-Based Forestry." In *Forest Policy and Politics in the Philippines: The Dynamics of Participatory Conservation,* edited by Peter Utting, 117–44. Quezon City: Ateneo de Manila University Press.

Baird, Ian. 2015. "Translocal Assemblages and the Circulation of the Concept of 'Indigenous Peoples' in Laos." *Political Geography* 46: 54–64.

Bankoff, Greg. 2004. "'The Tree as the Enemy of Man': Changing Attitudes to the Forests of the Philippines, 1565–1898." *Philippine Studies* 52, no. 3: 320–44.

Bankoff, Greg. 2013. "'Deep Forestry': Shapers of the Philippine Forests." *Environmental History* 18, no. 3: 523–56.

Banta, Jim E., Jerry W. Lee, Georgia Hodgkin, Zane Yi, Andrea Fanica, and Joan Sabate. 2018. "The Global Influence of the Seventh-day Adventist Church on Diet." *Religions* 9, no. 9: 251–76.

Barclay, Harold. 1986. *People without Government: An Anthropology of Anarchism.* London: Cienfuegos.

Barney, Keith. 2009. "Laos and the Making of a 'Relational' Resource Frontier." *Geographical Journal* 175, no. 2: 146–59.

Barrows, David Prescott. 1914. *A Decade of American Government in the Philippines, 1903–1913.* Chicago: World Book Company.

Barrows, David Prescott. 1924. *History of the Philippines.* New York: World Book Company.

Barrows, David Prescott. 1926. *History of the Philippines.* New York: World Book Company.

Barth, Fredrik. 1998. *Ethnic Groups and Boundaries: The Social Organization of Culture Difference.* Long Grove, IL: Waveland.

Bayuga, Rosy May M. 1989. *A Tagbanua Legacy: History of Palawan National Agricultural College.* Aborlan: Palawan National Agricultural College.

Beatty, Andrew. 2012. "The Tell-Tale Heart: Conversion and Emotion in Nias." *Ethnos* 77, no. 3: 295–320.

Bell, Joshua A., and Haidy Geismar. 2009. "Materialising Oceania: New Ethnographies of Things in Melanesia and Polynesia." *Australian Journal of Anthropology* 20, no. 1 (April): 3–27.

Bennett, Jane. 2009. *Vibrant Matter: A Political Ecology of Things.* Durham, NC: Duke University Press.

Berlant, Lauren. 1998. "Intimacy: A Special Issue." *Critical Inquiry* 24, no. 2: 281–88.

Bhattacharya, Tithi. 2017. "Social Reproduction Theory: Remapping Class, Recentering Oppression." For Work/Against Work: Debates on the Centrality of Work. Accessed January 2, 2022. https://onwork.edu.au/.

Bibal, Anna. 2021. "The Political Ecology of the Almaciga Forest Livelihood in Mount Mantalingahan, Palawan, Philippines." Master's thesis, University of the Philippines Los Baños.

Biermann, Christine, and Robert Anderson. 2017. "Conservation, Biopolitics, and the Governance of Life and Death." *Geography Compass* 11, no. 10: 1–13.

Blair, Emma, and James Robertson. 1973. *The Philippine Islands, 1493–1898.* Cleveland, OH: Arthur H. Clark.

Blanco, John D. 2009. *Frontier Constitutions. Christianity and Colonial Empire in the Nineteenth-Century Philippines.* Oakland, CA: University of California Press.

Bonsen, R., H. Marks, and J. Miedema, eds. 1990. *The Ambiguity of Rapprochement: Reflections of Anthropologists on Their Controversial Relationship with Missionaries.* Nijmegen, The Netherlands: Focaal (in cooperation with the NSAV).

Borup, Mads, Nik Brown, Kornelia Konrad, and Harro Van Lente. 2006. "The Sociology of Expectations in Science and Technology." *Technology Analysis and Strategic Management* 8, nos. 3–4: 285–98.

Bovensiepen, Judith, and Rosa Frederico. 2016. "Transformations of the Sacred in East Timor." *Comparative Studies in Society and History*, 58 no. 3: 664–93.

Brandon, Katrina, and Michael Wells. 1992. "Planning for People and Parks: Design Dilemmas." *World Development* 20, no. 4: 557–70.

Brookfield, Harold, and Christine Padoch. 1994. "Appreciating Agrodiversity: A Look at the Dynamism and Diversity of Indigenous Farming Practices." *Environment: Science and Policy for Sustainable Development* 36, no. 5: 6–45.

Brosius, J. Peter. 1999. "Analyses and Interventions: Anthropological Engagements with Environmentalism." *Current Anthropology* 40, no. 3: 277–310.

Brown, Elaine. 1991. "Tribal Peoples and Land Settlement: The Effects of Philippine Capitalist Development on the Palawan." PhD diss., State University of New York at Binghamton.

Bryant, Raymond. 2001. "Explaining State-Environmental NGO Relations in the Philippines and Indonesia." *Singapore Journal of Tropical Geography* 22, no. 1: 15–37.

Bryant, Raymond. 2002. "False Prophets? Mutant NGOs and Philippine Environmentalism." *Society & Natural Resources* 15, no. 7: 629–39.

Bryant, Raymond. 2008. *Nongovernmental Organizations in Environmental Struggles: Politics and the Making of Moral Capital in the Philippines.* New Haven, CT: Yale University Press.

Brosius, J. Peter. 1999 "Analyses and Interventions: Anthropological Engagements with Environmentalism." *Current Anthropology* 40, no. 3: 277–310.

Bull, Malcolm, and Keith Lockhart. 2006. *Seeking a Sanctuary: Seventh-day Adventism and the American Dream.* Bloomington: Indiana University Press.

Bureau of Forestry (Division of Insular Affairs, War Department). 1901. *Special Report of Captain George P. Ahern . . . in Charge of Forestry Bureau, Philippine Islands, Covering the Period from April, 1900, to July 30, 1901.* Washington: Government Printing Office,

Bureau of Forestry (Department of the Interior, The Philippine Commissions). 1904. *The Forest Manual (or the Forest Act No. 1148).* Manila: Bureau of Public Printing.

Bureau of Forestry. 1932. *Manual of Procedure.* Manila: Bureau of Public Printing.

Butler, Jonathan. 1986. "From Millerism to Seventh-day Adventism: 'Boundlessness to Consolidation.'" *Church History* 55, no. 1: 50–64.

Casumbal-Salazar, Melisa. S. 2015. "The Indeterminacy of the Philippine Indigenous Subject: Indigeneity, Temporality, and Cultural Governance." *Amerasia Journal* 41, no. 1: 74–94.

Cavanagh, Connor J. 2018. "Political Ecologies of Biopower: Diversity, Debates, and New Frontiers of Inquiry." *Journal of Political Ecology* 25, no. 1: 402–25.

Cepek, Michael. 2011. "Foucault in the Forest: Questioning Environmentality in Amazonia." *American Ethnologist* 38, no. 3: 501–15.

Chandler, David, and Julian Reid. 2019. *Becoming Indigenous: Governing Imaginaries in the Anthropocene*. Washington, DC: Rowman & Littlefield.

Chandler, David, and Julian Reid. 2020. "Becoming Indigenous: The 'Speculative Turn' in Anthropology and the (Re)Colonisation of Indigeneity." *Postcolonial Studies* 23, no. 4: 485–504.

Chapin, Mac. 2004. *A Challenge to Conservationists*. Washington, DC: World Watch Institute.

Chua, Liana. 2022. "'If God Is with Us, Who Can Be against Us?': Christianity, Cosmopolitics, and Living with Difference in Sarawak, Malaysian Borneo." *Current Anthropology* 63, no. 6: 714–36.

Clarke, Gerald. 1998. "Non-Governmental Organizations (NGOs) and Politics in the Developing World." *Political Studies* 46, no. 1: 36–52.

Clarke, Gerald. 2006. "Faith Matters: Faith-Based Organisations, Civil Society, and International Development." *Journal of International Development* 18: 835–48.

Clarke, John. 2014. "Conjunctures, Crises, and Cultures: Valuing Stuart Hall." *Focaal* 70: 113–22.

Clymer, Kenton J. 1980. "Religion and American Imperialism: Methodist Missionaries in the Philippine Islands, 1899–1913." *Pacific Historical Review* 49, no. 1: 29–50.

Clymer, Kenton J. 1982. "The Limits of Comity: Presbyterian-Baptist Relations in the Philippines, 1900–1925." *Missiology* 8, no. 2: 191–201.

Clymer, Kenton J. 1986. *Protestant Missionaries in the Philippines, 1898–1916: An Inquiry into the American Colonial Mentality*. Urbana: University of Illinois Press.

Cohen, Erik. 1989. "The Commercialization of Ethnic Crafts." *Journal of Design History* 2, nos. 2–3: 161–68.

Condominas, George. 1977. *We Have Eaten the Forest: The Story of a Montagnard Village in the Central Highlands of Vietnam*. New York: Farrar, Straus & Giroux.

Conklin, Beth A., and Laura Graham. 1995. "The Shifting Middle Ground: Amazonian Indians and Eco-politics." *American Anthropologist* 97, no. 4: 695–710.

Conklin, Harold. 1957. *Hanunoo Agriculture. A Report on an Integral System of Shifting Cultivation in the Philippines, Vol. 2*. Forestry Development Paper 12. Rome: Food & Agricultural Organization of the United Nations.

Conservation International. n.d. "Conservation Agreements: Making Conservation Attractive to Local Resource Users." Pdf accessed via www.conservation.org.

Conservation International 2010. CCA (Community Conservation Agreement) Progress Report. 2009–2010. *Agreement for Forest Protection of Mt Mantalingahan Protected Landscape, Marinsyawon, Bono-bono, Bildong*. Palawan, Philippines. Manila: Conservation International. Pdf obtained from Conservation International, Puerto Princesa City, Palawan, Philippines.

Conservation International. 2021. "Mount Mantalingahan. Our Shared Goal: No More Forest Loss." Accessed March 4, 2024. https://www.conservation.org/philippines/projects/mount-mantalingahan-protected-landscape.

Constitution of the Republic of the Philippines. 1987. Congress of the Philippines, Manila.

Cook, Ian. 2004. "Follow the Thing: Papaya." *Antipode* 36, no. 4: 642–64.

Cramb, Robert. A., Carol Colfer, Wolfram Dressler, Laungaramsri Pinkaew, Trang Le Quang, Mulyoutami Elok, Nancy Peluso, and Wadley Reed. 2009. "Swidden Transformations and Rural Livelihoods in Southeast Asia." *Human Ecology* 37, no. 3: 323–46.

Currie Peñaranda, Isabel, Silvia Otero-Bahamon, and Simón Uribe. 2021. "What Is the State Made Of? Coca, Roads, and the Materiality of State Formation in the Frontier." *World Development*. 141, no. 105395: 1–13.

Development Alternatives Incorporated (DAI). Conservation International, Orient Integrated Development Consultants, Rare and Tanggol Kalikasan. 2021. *Protect Wildlife Final Report*. USAID, Makati, Philippines. Accessed January 11, 2023. https://www.philchm.ph/wp-content/uploads/2021/11/Protect-Wildlife-Final-Report.2021-5-14_final.pdf.

Das, Veena, and Deborah Poole. 2004. "Anthropology in the Margins of the State." *PoLAR: Political and Legal Anthropology Review* 30, no. 1: 140–44.

Davis, Janet M. 2013. "Cockfight Nationalism: Blood Sport and the Moral Politics of American Empire and Nation Building." *American Quarterly* 65, no. 3: 549–74.

Dawdy, Shannon Lee. 2010. "Clockpunk Anthropology and the Ruins of Modernity." *Current Anthropology* 51, no. 6: 761–93.

De Beer, Jenne, and Melanie McDermott. 1989. *The Economic Value of Non-timber Forest Products in Southeast Asia: With Emphasis on Indonesia, Malaysia, and Thailand*. Gland: Netherlands Committee for IUCN.

De La Cadena, Marisel, and O. Starn, eds. 2007. *Indigenous Experience Today, Vol. 2*. Oxford: Berg.

Department of Environment and Natural Resources (DENR) 1993. Departmental Administrative Order No. 2 S, 1993). Rules and Regulations for the Identification, Delineation of Ancestral Land and Domain Claims, January 15, 1993, Manila, the Philippines.

Department of the Interior. 1910. *Report of the Governor General of the Philippines Islands*. Report of the Philippine Commission, June 30, 1910.

DiNovelli-Lang, Danielle, and Karen Hébert. 2018. "Ecological Labor." Society of Cultural Anthropology. Accessed July 26, 2018. https://culanth.org/fieldsights/ecological-labor.

Douglass, Michael. 2006. "Global Householding in Pacific Asia." *International Development Planning Review* 28, no. 4: 421.

Dove, Michael. 1983. "Theories of Swidden Agriculture, and the Political Economy of Ignorance." *Agroforestry Systems* 1: 85–99.

Dove, Michael. 1986. "The Practical Reason of Weeds in Indonesia: Peasant vs. State Views of Imperata and Chromolaena." *Human Ecology* 14, no. 2: 163–90.

Dove, Michael. 2021. *Bitter Shade: The Ecological Challenge of Human Consciousness*. New Haven, CT: Yale University Press.

Dressler, Wolfram H. 2009. *Old Thoughts in New Ideas: State Conservation Measures, Livelihood, and Development on Palawan Island, the Philippines*. Quezon City: Ateneo de Manila University Press.

Dressler, Wolfram H. 2014. "Green Governmentality and Swidden Decline on Palawan Island." *Transactions of the Institute of British Geographers* 39, no. 2: 250–64.

Dressler, Wolfram H. 2017. "Contesting Moral Capital in the Economy of Expectations of an Extractive Frontier." *Annals of the American Association of Geographers* 107, no. 3: 647–65.

Dressler, Wolfram H. 2019. "Governed from Above, Below, and Dammed in Between: The Biopolitics of (Un)Making Life and Livelihood in the Philippine Uplands." *Political Geography* 73: 123–37.

Dressler, Wolfram H. 2021. "Defending Lands and Forests: NGO Histories, Everyday Struggles, and Extraordinary Violence in the Philippines." *Critical Asian Studies* 52, no. 3: 1–32.

Dressler, Wolfram H., Bram Büscher, Michael Schoon, Dan Brockington, Tanya Hayes, Christian Kull, James McCarthy, and Krishna Shrestha. 2010. "From Hope to Crisis and Back Again? A Critical History of the Global CBNRM Narrative." *Environmental Conservation* 37, no. 1: 5–15.

Dressler, Wolfram H., and Eli Guieb. 2015. "Violent Enclosures, Violated Livelihoods: Environmental and Military Territoriality in a Philippine Frontier." *Journal of Peasant Studies* 42, no. 2: 323–45.

Dressler, Wolfram H., and Fabinyi Michael. 2011. "Farmer Gone Fish'n? Swidden Decline and the Rise of Grouper Fishing on Palawan Island, the Philippines." *Journal of Agrarian Change* 11, no. 4: 536–55.

Dressler, Wolfram H., and Juan Pulhin. 2010. "The Shifting Ground of Swidden Agriculture on Palawan Island, the Philippines." *Agriculture and Human Values* 27, no. 4: 445–59.

Dressler, Wolfram H., and Sarah Turner. 2008. "The Persistence of Social Differentiation in the Philippine Uplands." *Journal of Development Studies* 44, no. 10: 1450–73.

Dressler, Wolfram H., Will Smith, Christian A. Kull, Rachel Carmenta, and Juan M. Pulhin. 2021. "Recalibrating Burdens of Blame: Anti-Swidden Politics and Green Governance in the Philippine Uplands." *Geoforum* 124: 348–59.

Dressler, Wolfram, Will Smith, and Marvin Montefrio. 2018. "Ungovernable? The Vital Natures of Swidden Assemblages in an Upland Frontier." *Journal of Rural Studies* 61: 343–54.

Dumont, Jean Paul. 1992. "Ideas of Philippine Violence: Assertions, Negations, and Narrations." In *The Paths to Domination, Resistance, and Terror*, edited by Carolyn Nordstrom and Joann Martin, 133–53. Oakland: University of California Press.

Duncan, Christopher. 2008. *Civilizing the Margins: Southeast Asian Government Policies for the Development of Minorities*. Singapore: National University of Singapore Press.

Dunch, Ryan. 2002. "Beyond Cultural Imperialism: Cultural Theory, Christian Missions, and Global Modernity." *History and Theory* 41, no. 3: 301–25.

Eder, James. 1987. *On the Road to Tribal Extinction: Depopulation, Deculturation, and Adaptive Well-being among the Batak of the Philippines*. Berkeley: University of California Press.

Eder, James. 1999. *A Generation Later: Household Strategies and Economic Change in the Rural Philippines*. Honolulu: University of Hawai'i Press.

Eder, James. 1999. *A Generation Later: Household Strategies and Economic Change in the Rural Philippines*. Honolulu: University of Hawai'i Press.

Eder, James. 2010. "Ethnic Differences, Islamic Consciousness, and Muslim Social Integration in the Philippines." *Journal of Muslim Minority Affairs* 30, no. 3: 317–32.

Eder, James, and Janet Fernandez. 1996. *Palawan at the Crossroads*. Quezon City: Ateneo de Manila University Press.

Eder, James, and Thomas Mckenna. 2004. "Minorities in the Philippines." In *Civilizing the Margins: Southeast Asian Government Policies for the Development of Minorities*, edited by Christopher Duncan, 56–85. Ithaca, NY: Cornell University Press.

Eilenberg, Michael. 2014. "Frontier Constellations: Agrarian Expansion and Sovereignty on the Indonesian-Malaysian Border." *Journal of Peasant Studies* 41, no. 2: 157–82.

Ellis, Frank. 1993. *Peasant Economics*. Cambridge: Cambridge University Press.

Escobar, Arturo. 2020. *Pluriversal Politics*. Durham, NC: Duke University Press.

Fabinyi, Michael, Wolfram H. Dressler, and Michael Pido. 2019. "Access to Fisheries in the Maritime Frontier of Palawan Province, Philippines." *Singapore Journal of Tropical Geography* 40, no. 1: 92–110.

Farrier, David. 2019. *Anthropocene Poetics: Deep Time, Sacrifice Zones, and Extinction.* Minneapolis: University of Minnesota Press.

Ferguson, James. 1994. *The Anti-Politics Machine: "Development," Depoliticization, and Bureaucratic Power in Lesotho.* Minneapolis: University of Minnesota Press.

Ferguson, James. 2006. *Global Shadows: Africa in the Neoliberal World Order.* Durham, NC: Duke University Press.

Fernandez, Bina. 2018. "Dispossession and the Depletion of Social Reproduction." *Antipode* 50, no. 1: 142–63.

Fernandez, Gil. 1990. *Light Dawns over Asia: Adventism's Story in the Far Eastern Division: 1888–1988.* Cavite: Adventist International Institute of Advanced Studies Publications.

Fisher, William. 1997. "Doing Good? The Politics and Anti-Politics of NGO Practices." *Annual Review of Anthropology* 26: 439–64.

Fletcher, Robert, Wolfram H. Dressler, Anderson Zachary, and Büscher Bram. 2019. "Natural Capital Must Be Defended: Green Growth as Neoliberal Biopolitics." *Journal of Peasant Studies* 46, no. 5: 1068–95.

Fletcher, Robert. 2017. "Environmentality Unbound: Multiple Governmentalities in Environmental Politics." *Geoforum* 85: 311–15.

Foster, Robert. 2006. "Tracking Globalization: Commodities, and Value in Motion." In *Handbook of Material Culture*, edited by Christopher Tilley, Webb Keane, Susan Kuechler, Mike Rowlands, and Patricia Spyer, 285–302. Thousand Oaks, CA: Sage.

Foucault, Michel. 1978. *The History of Sexuality.* New York: Vintage.

Foucault, Michel. 1983. "The Subject and Power." In *Michel Foucault: Beyond Structuralism and Hermeneutics*, edited by Dreyfus Hubert and Paul Rabinow, 208–26. Chicago: University of Chicago Press.

Foucault, Michel. 2003. *Society Must Be Defended: Lectures at the College de France, 1975–76.* Edited by Mauro Bertani and Alessandro Fontana. New York: Picador.

Foucault, Michel. 2007. *Security, Territory, Population: Lectures at the Collège de France, 1977–78.* Amsterdam: Springer.

Fox, Robert. 1954. "Tagbanua Religion and Society." PhD diss., University of Chicago.

Gell, Alfred. 1998. *Art and Agency: An Anthropological Theory.* Oxford: Clarendon.

Gordon, Colin. 1991. "Governmental Rationality: An Introduction." In *The Foucault Effect: Studies in Governmentality*, edited by Graham Burchell, Colin Gordon, and Peter Miller, 1–52. Chicago: University of Chicago Press.

Graeber, David. 2001. *Toward an Anthropological Theory of Value: The False Coin of Our Own Dreams.* Palgrave Macmillan.

Greenleaf, Maron. 2021. "Beneficiaries of Forest Carbon: Precarious Inclusion in the Brazilian Amazon." *American Anthropologist* 123, no. 2: 305–17.

Grimes, Kimberly, and Lynne Milgram. 2000. *Artisans and Co-operatives: Developing Alternative Trade for the Global Economy.* Tucson: University of Arizona Press.

Gospel Ministries International TV. "Kent and Leonda George Testimony." YouTube video, 38:02, August 2, 2021. https://www.youtube.com/watch?v=PhhkvW8eyx8.

Guarasci, Bridget, and Eleana Kim. 2022. "Introduction: Ecologies of War." Editors' Forum: Theorizing the Contemporary, Society for Cultural Anthropology. Accessed February 22, 2023. https://culanth.org/fieldsights/introduction-ecologies-of-war.

Guerrero, Abraham. 2010. "Structure and Mission Effectiveness: A Study Focused on Seventh-day Adventist Missions to Unreached People Groups between 1980 and 2010." PhD diss., Andrews University.

Guieb, Eli R., III. 1999. "Reasserting Indigenous Spaces in a Tagbanua Text: A Case of Dagat Ninuno (Ancestral Water Resource Claims) in the Philippines." In "On Power, Spaces, and Titles," special issue, *Lundayan Journal* 1: 23–33.

Hall, Rosalie Arcala, and Imelda Deinla. 2021. "Shifts in the Humanitarian Space? Examining NGO–Military Engagements during the 2017 Crisis in Marawi, Philippines." *Asian Politics & Policy* 13, no. 3: 349–65.

Hall, Stuart. 2011. "The Neo-Liberal Revolution." *Cultural Studies* 25, no. 6:705–28.

Hall, Stuart, and Doreen Massey. 2012. "Interpreting the Crisis." In *The Neoliberal Crisis,* edited by Jonathan Rutherford and Sally Davison, 55–70. London: Lawrence & Wishart.

Hefner, Robert W. 1993. *Conversion to Christianity: Historical and Anthropological Perspectives on a Great Transformation.* Oakland: University of California Press.

Herzfeld, Michael. 2002. "The Absence Presence: Discourses of Crypto-Colonialism." *South Atlantic Quarterly* 101, no. 4: 899–926.

Hilhorst, Dorothea. 2000. *Records and Reputations.* Quezon City: Ateneo de Manila University Press.

Hirtz, Frank. 2003. "It Takes Modern Means to Be Traditional: On Recognizing Indigenous Cultural Communities in the Philippines." *Development and Change* 34, no. 5: 887–914.

Hobbes, Marieke. 2000. "Pala'wan Managing Their Forest: The Case of the Pala'wan of Saray on Southern Palawan Island, the Philippines." Master's thesis, Leiden University.

Hobsbawm, Eric, and Terence Ranger. 2012. *The Invention of Tradition.* Cambridge: Cambridge University Press.

Holden, William. 2012. "Ecclesial Opposition to Large-Scale Mining on Samar: Neoliberalism Meets the Church of the Poor in a Wounded Land." *Religions* 3, no. 3: 833–61.

Hoskins, Janet. 1998. *Biographical Objects: How Things Tell the Stories of People's Lives.* New York: Routledge.

Hoskins, Janet. 2006. "Agency, Biography, and Objects." In *Handbook of Material Culture,* edited by Christopher Tilley, Webb Keane, Susanne Küchler, Mike Rowlands, and Patricia Spyer, 74–84. Thousand Oaks, CA: Sage.

Howell, Brian. 2009. "Moving Mountains: Protestant Christianity and the Spiritual Landscape of Northern Luzon." *Anthropological Forum* 19, no. 3: 253–69.

Hubinger, Vaclav. 1997. "Anthropology and Modernity." *International Social Science Journal* 49, no. 154: 527–36.

Hurst, Philip. 1992. "Philippines." In *Rainforest Politics: Ecological Destruction in Southeast Asia,* edited by Philip Hurst, 162–205. London: Zed.

Hutterer, Karl. 1978. "Dean Worcester and Philippine Anthropology." *Philippine Quarterly of Culture and Society* 6, no. 3: 125–56.

Igoe, James. 2005. "Global Indigenism and Spaceship Earth: Convergence, Space, and Re-entry Friction." *Globalizations* 2, no. 3: 377–90.

Ileto, Reynaldo Clemeña. 1979. *Pasyon and Revolution: Popular Movements in the Philippines, 1840–1910.* Quezon City: Ateneo de Manila University Press.

Indigenous Peoples Rights Act. 1997. Republic Act No. 8371, Congress of the Philippines.

Iskander, Dalia. 2021. *The Power of Parasites: Malaria as (un) Conscious Strategy.* Amsterdam: Springer Nature.

Johnston, Alison. 2014. *Is the Sacred for Sale: Tourism and Indigenous Peoples.* London: Routledge.

Jones, Arun. 2002. "A View from the Mountains: Episcopal Missionary Depictions of the Igorot of Northern Luzon, the Philippines, 1903–1916." *Anglican and Episcopal History* 71, no. 3: 380–410.

Jordana y Morera, Ramon. 1879. *Memoria Sobre la Producción de Los Montes Públicos de Filipinas en los Años Económicos de 1873–74 y 1875–76*. Madrid: Imprenta de Ramon Mereno y Ricardo Rojas.

Joslin, Audry. 2022. "Labor as a Linchpin in Ecosystem Services Conservation: Appropriating Value from Collective Institutions?" *Capitalism Nature Socialism* 33, no. 1: 90–110.

Kalaw, Maximo. 1919. "Recent Policy towards the Non-Christian People of the Philippines." *Journal of International Relations* 10, no. 1: 1–12.

Katz, Cindi. 2004. *Growing Up Global: Economic Restructuring and Children's Everyday Lives*. Minneapolis: University of Minnesota Press.

Keane, Webb. 2007. *Christian Moderns: Freedom and Fetish in the Mission Encounter*. Berkeley: University of California Press.

Kerkvliet, Ben. 2009. "Everyday Politics in Peasant Societies (and Ours)." *Journal of Peasant Studies* 36, no. 1: 227–43.

Kofman, Elenore. 2014. "Gendered Migrations, Social Reproduction, and the Household in Europe." *Dialectical Anthropology* 38, no. 1: 79–94.

Kopytoff, Igor. 1986. "The Cultural Biography of Things: Commoditization as Process." *The Social Life of Things: Commodities in Cultural Perspective* 68: 70–73.

Korf, Benedikt, and Timothy Raeymaekers. 2013. "Introduction: Border, Frontier, and the Geography of Rule at the Margins of the State." In *Violence on the Margins*, edited by Korf Benedikt, 3–27. New York: Palgrave Macmillan.

Kuper, Adam. 2003. "The Return of the Native." *Current Anthropology* 44, no. 3: 389–402.

Land, Gary. 2015. *Historical Dictionary of the Seventh-day Adventists*. 2nd ed. New York: Rowman & Littlefield.

Laurie, Clayton. 1989. "The Philippine Scouts: America's Colonial Army, 1899–1913." *Philippine Studies* 37, no. 2: 174–91.

Leach, Edmund. 1977. *Political Systems of Highland Burma: A Study of Kachin Social Structure*. London: Taylor & Francis.

Lemke, Thomas. 2011. "Critique and Experience in Foucault." *Theory, Culture, and Society* 28, no. 4: 26–48.

Lemke, Thomas, M. Casper, and L. J. Moore. 2011. *Biopolitics*. New York: New York University Press.

Lerner, Steve. 2012. *Sacrifice Zones: The Front Lines of Toxic Chemical Exposure in the United States*. Cambridge, MA: MIT Press.

Li, Tania Murray. 1999. *Transforming the Indonesian uplands: Marginality, Power, and Production*. Canada: Harwood Academic Publishers.

Li, Tania Murray. 2000. "Articulating Indigenous Identity in Indonesia: Resource Politics and the Tribal Slot." *Comparative Studies in Society and History* 42, no. 1: 149–79.

Li, Tania Murray. 2005. "Beyond 'the State' and Failed Schemes." *American Anthropologist* 107, no. 3: 383–94.

Li, Tania Murray. 2007. *The Will to Improve: Governmentality, Development, and the Practice of Politics*. Durham, NC: Duke University Press.

Li, Tania Murray. 2014. *Land's End: Capitalist Relations on an Indigenous Frontier*. Durham, NC: Duke University Press.

Li, Tania Murray. 2016. "Governing Rural Indonesia: Convergence on the Project System." *Critical Policy Studies* 10, no. 1: 79–94.

Li, Tania Murray. 2018. "Fixing Non-market Subjects: Governing Land and Population in the Global South." In *Governing Practices: Neoliberalism, Governmentality, and the Ethnographic Imaginary,* edited by Michelle Brady and Randy Lippert, 80–102. Toronto: University of Toronto Press.

Li, Tania Murray. 2019. "Problematizing the Project System: Rural Development in Indonesia." In *The Projectification of the Public Sector*, 56–74. New York: Routledge.

Lund, Christian. 2021. *Nine-Tenths of the Law: Enduring Dispossession in Indonesia.* New Haven, CT: Yale University Press.

Lynch, Owen. 1982. "Native Title, Private Right, and Tribal Land Law: An Introductory Survey." *Philippine Law Journal* 57: 268–305.

McCoy, Alfred. 1982. "Baylan: Animist Religion and Philippine Peasant Ideology." *Philippine Quarterly of Culture and Society* 10: 141–94.

Macdonald, Charles. 1997. "Cleansing the Earth: The 'Pänggaris' Ceremony in Palawan." *Philippine Studies* 45, no. 3: 408–22.

Macdonald, Charles. 1992a. "Protestant Missionaries and Palawan Natives: Dialogue, Conflict, or Misunderstanding?" *Journal of the Anthropological Society of Oxford* 23, no. 2: 127–37.

Macdonald, Charles. 1992b. "Invoking the Spirits in Palawan: Ethnography and Pragmatics." In *Sociolinguistics Today*, edited by Kingsley Bolton and Kwok Helen, 244–60. London: Routledge.

Macdonald, Charles. 2004. "Folk Catholicism and Pre-Spanish Religions in the Philippines." *Philippine Studies*, 52, no. 1:78–93.

Macdonald, Charles. 2007. *Uncultural Behavior: An Anthropological Investigation of Suicide in the Southern Philippines.* Honolulu: University of Hawai'i Press.

Macdonald, Charles. 2009. "The Anthropology of Anarchy." Paper No. 35, Occasional Papers of the School of Social Science, Institute for Advanced Study. Accessed March 4, 2024. https://theanarchistlibrary.org/library/charles-j-h-macdonald-the-anthropology-of-anarchy.

Macdonald, Charles. 2011a. "Primitive Anarchs: Anarchism and the Anthropological Imagination." *Social Evolution & History* 10, no. 2: 67–86.

Macdonald, Charles. 2011b. "Kinship and Fellowship among the Palawan." In *Anarchic Solidarity: Autonomy, Equality, and Fellowship in Southeast Asia*, edited by Thomas Gibson and Silander Kenneth, 152–75. New Haven, CT: Yale University Press.

Macdonald, Charles. 2012. "The Anthropology of Anarchy." *Indian Journal of Human Development* 6, no. 1: 49–66.

Macdonald, Charles. 2013. "The Filipino as Libertarian: Contemporary Implications of Anarchism." *Philippine Studies: Historical and Ethnographic Viewpoints* 61, no. 4: 413–36.

Malkki, Lisa. 1992. "National Geographic: The Rooting of Peoples and the Territorialization of National Identity among Scholars and Refugees." *Cultural Anthropology* 7, no. 1: 24–44.

Massarella, Kate, Susannah Sallu, Jonathan Ensor, and Rob Marchant. 2018. "REDD+, Hype, Hope, and Disappointment: The Dynamics of Expectations in Conservation and Development Pilot Projects." *World Development* 109: 375–85.

Massey, Doreen. 1993. "Questions of Locality." *Geography* 78, no. 2: 142–49.

Massey, Doreen. 1994. *Space, Place, and Gender*. Minneapolis: University of Minnesota Press.

McDermott, Melanie Hughes. 2000. *Boundaries and Pathways: Indigenous Identity, Ancestral Domain, and Forest Use in Palawan, the Philippines*. Berkeley: University of California.

McDermott, Melanie Hughes. 2001. "Invoking Community: Indigenous People and Ancestral Domain in Palawan, the Philippines." In *Community and the Environment: Ethnicity, Gender, and the State in Community-based Conservation*, edited by Arun Agrawal and Clark Gibson, 32–62. New Brunswick, NJ: Rutgers University Press.

McDougall, D. 2020. "Beyond Rupture: Christian Culture in the Pacific." *Australian Journal of Anthropology* 31, no. 2: 203–9.

McElwee, Pamela. 2016. *Forests Are Gold: Trees, People, and Environmental Rule in Vietnam*. Seattle: University of Washington Press.

McKay, Deirdre. 2006. "Rethinking Indigenous Place: Igorot Identity and Locality in the Philippines." *Australian Journal of Anthropology* 17, no. 3: 291–306.

McKay, Deirdre, and Padma Perez. 2018. "Plastic Masculinity: How Everyday Objects in Plastic Suggest Men Could Be Otherwise." *Journal of Material Culture* 23, no. 2: 169–86.

McKenna, Thomas. 1996. "National Ideas and Rank-and-File Experience in the Muslim Separatist Movement in the Philippines." *Critique of Anthropology* 16, no. 3: 229–55.

Merlan, Francesca. 2009. "Indigeneity: Global and Local." *Current Anthropology* 50, no. 3: 303–33.

Milgram, Lynee. 2001. "Operationalising Microfinance: Women and Craftwork in Ifugao, Upland Philippines." *Human Organization* 60, no. 3: 212–24.

Miller, Daniel. 2005. "Materiality: An Introduction." In *Materiality*, edited by Miller Daniel, 1–50. Durham, NC: Duke University Press.

Milne, Sarah. 2009. "Global Ideas, Local Realities: The Political Ecology of Payments for Biodiversity Conservation Services in Cambodia." PhD diss., University of Cambridge.

Mitlin, Diana, Sam Hickey, and Anthony Bebbington. 2007. "Reclaiming Development? NGOs and the Challenge of Alternatives." *World Development* 35, no. 10: 1699–720.

Mojares, Resil. 2013. "Jose Rizal in the World of German Anthropology." *Philippine Quarterly of Culture and Society* 41, nos. 3–4: 163–94.

Montefrio, Marvin. 2017. "Land Control Dynamics and Social-Ecological Transformations in Upland Philippines." *Journal of Peasant Studies* 44, no. 4: 796–816.

Montefrio, Marvin, and Wolfram H. Dressler. 2016. "The Green Economy and Constructions of the 'Idle' and 'Unproductive' Uplands in the Philippines." *World Development* 79: 114–26.

Mount Mantalingahan Protected Landscape (MMPLA). 2010. "Mount Mantalingahan Protected Landscape Management Plan." European Union, PCSDs, Republic of the Philippines.

Murti, Radhika, and Camille Buyck, eds. 2014. *Protected Areas for Disaster Risk Reduction and Climate Change Adaptation*. Gland, Switzerland: International Union for the Conservation of Nature.

Nano, Jose. 1939. "Kaingin Laws and Penalties in the Philippines." *Philippines Journal of Forestry* 2, no. 2: 87–92.

Neimark, Ben, Sango Mahanty, Wolfram Dressler, and Christina Hicks. 2020. "Not Just Participation: The Rise of the Eco-Precariat in the Green Economy." *Antipode* 52, no. 2: 496–521.

Nelson, Nici, and Susan Wright. 1995. "Introduction: Participation and Power." In *Power and Participatory Development,* edited by Nici Nelson and Susan Wright, 1–18. London: Intermediate Technology.

Netting, Robert. 1993. *Smallholders, Householders: Farm Families and the Ecology of Intensive, Sustainable Agriculture.* Stanford, CA: Stanford University Press.

Niesten, Eduard, Patricia Zurita, and Sarah Banks. 2010. "Conservation Agreements as a Tool to Generate Direct Incentives for Biodiversity Conservation." *Biodiversity* 11, nos. 1–2: 5–8.

Niezen, Ronald. 2003. *The Origins of Indigenism.* Oakland: University of California Press.

Noblejas, Antonio, and Edilerto Noblejas. 1992. *Registration of Land Titles and Deeds in the Philippines.* Manila: Rex.

Novellino, Dario. 2001. "Pälawan Attitudes towards Illness." *Philippine Studies* 49, no. 1: 78–93.

Novellino, Dario. 2007. "Cycles of Politics and Cycles of Nature: Permanent Crisis in the Uplands of Palawan." In *Modern Crisis and Traditional Strategies: Local Ecological Knowledge in Island Southeast Asia,* edited by Roy Ellen, 185–219. New York: Berghahn.

Novellino, Dario. 2011. "Toward a 'Common Logic of Procurement': Unravelling the Foraging-Farming Interface on Palawan Island (the Philippines)." In *Why Cultivate? Anthropological and Archaeological Approaches to Foraging-Farming Transitions in Southeast Asia,* edited by Graeme Barker and Monica Janowski, 105–19. Cambridge: McDonald Institute Monographs.

NTFP-EP. 2008. *The Tingkep and Other Crafts of the Pala'wan.* Non-timber Forest Product Exchange Programme for South and Southeast Asia on Behalf of the Pala'wan Community and NATRIPAL. Diliman, Quezon City, The Philippines.

O'Brien, William. 2002. "The Nature of Shifting Cultivation: Stories of Harmony, Degradation, and Redemption." *Human Ecology* 30, no. 4: 483–502.

Ocampo, Nilo. 1985. *Katutubo, Muslim, Christian: Palawan, 1621–1901.* Cologne: Karapatang Sipi ng mga Tagapagpalathala, ng May-akda, at ng Bahay-Saliksikan ng Kasaysayan (BAKAS), 157.

Ocampo, Nilo. 1996. "A History of Palawan." In *Palawan at the Crossroads,* edited by James Eder and Janet Fernandez, 23–37. Quezon City: Ateneo de Manila University Press.

Olofson, Harold. 1980. "Swidden and Kaingin among the Southern Tagalog: A Problem in Philippine Upland Ethno-Agriculture." *Philippine Quarterly of Culture and Society* 8, nos. 2–3: 168–80.

O'Malley, Pat, Lorna Weir, and Clifford Shearing. 1997. "Governmentality, Criticism, Politics." *Economy and Society* 26, no. 4: 501–17.

Palawan Council for Sustainable Development (PCSD). 1992. "Strategic Environmental Plan." Republic Act 7611, Republic of the Philippines.

Palawan Tropical Forestry Protection Programme (PTFPP). 1997. "PTFPP Third Quarter Progress Report." Puerto Princesa City, Palawan Island, the Philippines.

Paredes, Oona. 2006. "True Believers: Higaunon and Manobo Evangelical Protestant Conversion in Historical and Anthropological Perspective." *Philippine Studies* 54, no. 4: 521–59.

Paredes, Oona. 2013. *A Mountain of Difference. The Lumad in Early Colonial Mindanao.* Ithaca, NY: Cornell University Press.

Parks, Thomas., 2008. "The Rise and Fall of Donor Funding for Advocacy NGOs: Understanding the Impact." *Development in Practice* 18, no. 2: 213–22.

PD 705 (Presidential Decree 705). 1975. The revised forestry code. Congress of the Philippines, Manila.

Peluso, Nancy Lee. 1990. *Rich Forests, Poor People: Resource Control and Resistance in Java*. Oakland: University of California Press.

Peluso, Nancy Lee, and Agus Budi Purwanto. 2018. "The Remittance Forest: Turning Mobile Labor into Agrarian Capital." *Singapore Journal of Tropical Geography* 39, no. 1: 6–36.

Peterson, Jean Treloggen. 1978. "Hunter-Gatherer/Farmer Exchange." *American Anthropologist* 80, no. 2: 335–51.

Perez, Padmapani, and the Philippine ICCA Consortium (Bukluran). 2018. "Living with the Problem of National Parks: Indigenous Critique of Philippine Environmental Policy." *Thesis Eleven* 145, no. 1: 58–76.

Philippine Commission. 1901. *Report of the United States Philippine Commission to the Secretary of War from December 1, 1900, to October 15, 1901, Vol. 1*. 4 vols. Washington, DC: Government Printing Office.

Philippine Commission. 1903. *Census of the Philippine Islands: Taken under the Direction of the Philippine Commission in the Year 1903, Vol. 2*. 4 vols. Washington, DC: Government Printing Office, 1905).

Philippine Council for Agriculture, Forestry, and Natural Resources Research and Development (PCARRD). 1991. *A Beginner's Sourcebook on Philippine Rattan*. Laguna: Los Baños.

Pinchot, Gifford. 1903. "The Forester and the Lumberman." *Forestry and Irrigation* (April): 176–78.

Pinto, Femy. 1999. "Contesting Frontier Lands in Palawan, Philippines: Strategies of Indigenous Peoples for Community Development and Ancestral Domain Management." Master's thesis, Clark University.

Population Center Foundation. 1980. *Kaingineros: The Philippine Boat People, National Task Force*. Makati: Population Center Foundation.

Povinelli, Elizabeth. A. 2002. *The Cunning of Recognition*. Durham, NC: Duke University Press.

Priit, J. Vesilind 2002. "The Philippines—Hotspots: Preserving Pieces of a Fragile Biosphere." *National Geographic* 202, no. 1: 62.

PTFPP. 2000. *Catchment Management Plan, Tamlang Catchment, Brooke's Point*. Puerto Princesa City, Palawan Island, the Philippines.

PTFPP. 2002a. *Catchment Management Plan for Tamlang, Brooke's Point*. Puerto Princesa City, Palawan Island, the Philippines.

PTFPP. 2002b. *Inogbong Catchment Management Plan. Vols. 1 and 2. Description, Evaluation, and Management Plan*. Puerto Princesa City, Palawan Island, the Philippines.

PTFPP. 2002c. *Summary Report on the Generation of the First-Generation Community and Livelihood Micro-Projects*. By Jovy Cajegas, Community and Livelihood Department. Puerto Princesa City, Palawan Island, the Philippines.

PTFPP. 2002d. *PTFPP Community Micro-project Implementation*. Puerto Princesa City, Palawan Island, the Philippines.

PTFPP. 2002e. *Proceedings of Community Consultations: Presentation of Tribal Learning Programme fir IPs at Sitio of Dao, Cogon-Cogon, and Kilala of Tamlang Catchment, Brooke's Point, Palawan*. Palawan Tropical Forestry Protection Programme, Brookes LGU and the Palawan Council for Sustainable Development Staff (PCSDs).

Pulhin, Juan. 2000. "Community Forestry in the Philippines: Paradoxes and Perspectives in Development Practice." Paper presented at Eighth Biennial Conference of the International Association for the Study of Common Property (IASCP), Bloomington, Indiana, May 31–June 4, 2000.

Rabinow, Paul, and Nicholas Rose. 2006. "Biopower Today." *BioSocieties* 1, no. 2: 195–217.

Rafael, Vicente. 2000. *Contracting Colonialism: Translation and Christian Conversion in Tagalog Society under Early Spanish Rule.* Durham, NC: Duke University Press.

Rasmussen, Mattias, and Lund Christian. 2018. "Reconfiguring Frontier Spaces: The Territorialization of Resource Control." *World Development* 101: 388–99.

Raymond, Joseph. 2008. "Colonial Apostles: A Discourse on Syncretism and the Early American Protestant Missions in the Philippines." *Asia-Pacific Social Science Review* 8, no. 1: 140–46.

Razal, Ramon, Aeron Maralit, Norlita Colili, Loreta Alsa, and Ruth Canlas. 2013. *Value Chain Study on Palawan Almaciga Resin.* Manila: NTFP-EP.

Revel-Macdonald, Nicole. 1977. "The Magic Doll." In *Filipino Heritage. The Making of a Nation. The Age of Trade and Contacts, Visitors from across many Seas*, vol. 3, 570–73. Manila: Lahing Pilipino.

Revel, Nicole. 1990a. *Fleurs de Paroles: Histoire Naturelle Palawan. I. Les Dons de Nägsalad.* Leuven: Peeters.

Revel, Nicole. 1990b. *Fleurs de Paroles: Histoire Naturelle Palawan. II La Maitrise d'un Savoir et L'Art D'une Relation.* Leuven: Peeters.

Revel, Nicole. 2011. "The Gifts of the Weaver and Their Becoming at the Turn of the 21st Century." *Agham Tao* 20:49–67.

Revel, Nicole, Jennifer Gage, and Patricia Railing. 1998. "As if in a Dream . . . ' Epics and Shamanism among Hunters. Palawan Island, the Philippines." *Diogenes* 46, no. 181: 7–30.

Revel, Nicole, Xhauflair Hermine, and Norlita Colili. 2017. "Childhood in Pala'wan Highlands Forest, the Känakan (Philippines)." *AnthropoChildren* 7, no. 7: 1–27.

Rosales, Christian. 2020. "Tigi: Justice and Indigenous Citizenship among the Iraya Mangyan in Mindoro." *Social Science Diliman* 16, no. 2:66–103.

Rosales, Christian. 2021. "Pigs and Ritual-Hunting among the Highland Tau-Buhid in Mounts Iglit-Baco Natural Park, Philippines." *Anthropozoologica* 56, no. 9: 137–51.

Roth, Dennis. 1983. "Philippine Forests and Forestry, 1565–1920." In *Global Deforestation and the Nineteenth-Century World Economy*, edited by Richard Tucker and J. F. Richards, 41–55. Durham, NC: Duke University Press.

Rubis, June, and Noah Theriault. 2020. "Concealing Protocols: Conservation, Indigenous Survivance, and the Dilemmas of Visibility." *Social and Cultural Geography* 21, no. 7: 962–84.

Saguin, Kristian. 2016. "Blue Revolution in a Commodity Frontier: Ecologies of Aquaculture and Agrarian Change in Laguna Lake, Philippines." *Journal of Agrarian Change* 16, no. 4: 571–93.

Schiller, Anne. 2009. "On the Catholic Church and Indigenous Identities: Notes from Indonesian Borneo." *Culture and Religion.* 10, no. 3: 279–95.

Scott, Geoffrey. 1979. "Kaingin Management in the Republic of the Philippines." *Philippine Geographical Journal* 23, no. 2: 58–73.

Scott, James. 1977. *The Moral Economy of the Peasant.* New Haven, CT: Yale University Press.

Scott, James. 1986. "Everyday Forms of Peasant Resistance." *Journal of Peasant Studies* 13, no. 2: 5–35.

Scott, James. 2008. *Weapons of the Weak: Everyday Forms of Peasant Resistance.* New Haven, CT: Yale University Press

Scott, James. 2008. *Seeing Like a State: How Certain Schemes to Improve the Human Condition Have Failed*. New Haven, CT: Yale University Press

Scott, James. 2009. *The Art of Not Being Governed*. New Haven, CT: Yale University Press.

Scott, William Henry. 1974. *The Discovery of the Igorots: Spanish Contacts with the Pagans of Northern Luzon*. Quezon City: New Day.

Semple, Rhonda. 2017. "Making Missions through (Re-) Making Children: Non-Kin Domestic Intimacy in the London Missionary Society's Work in Late Nineteenth-Century North India." In *Creating Religious Childhoods in Anglo-World and British Colonial Contexts, 1800–1950*, edited by Hugh Morrison and Mary Clare Martin, 35–52. London: Routledge.

Shepherd, Christopher, and Andrew McWilliam. 2011. "Ethnography, Agency, and Materiality: Anthropological Perspectives on Rice Development in East Timor." *East Asia Science, Technology, and Society: An International Journal* 5, no. 2: 189–215.

Simpson, Fergus O'Leary, and Lorenzo Pellegrini. 2022. "Conservation, Extraction, and Social Contracts at a Violent Frontier: Evidence from Eastern DRC's Itombwe Nature Reserve." *Political Geography* 92: 1–11.

Smith, Will. 2015. "Swidden, Climate, and Culture: Negotiating 'Double Exposure' amongst Indigenous Households on Palawan Island, the Philippines." PhD diss., University of Queensland.

Smith, Will. 2018. "Weather from Incest: The Politics of Indigenous Climate Change Knowledge on Palawan Island, the Philippines." *Australian Journal of Anthropology* 29, no. 3: 265–81.

Smith, Will. 2020. "Beyond Loving Nature: Affective Conservation and Human-Pig Violence in the Philippines." *Ethnos*, 87, no. 5:1–19.

Smith, Will. 2021. *Mountains of Blame: Climate and Culpability in the Philippine Uplands*. Seattle: University of Washington Press.

Smith, Will, and Wolfram H. Dressler. 2017. "Rooted in Place? The Coproduction of Knowledge and Space in Agroforestry Assemblages." *Annals of the Association of American Geographers* 107, no. 4: 897–914.

Smith, Will, and Wolfram H. Dressler. 2019. "Governing Vulnerability: The Biopolitics of Conservation and Climate in Upland Southeast Asia." *Political Geography*, no. 72: 76–86.

Smith, Will, and Wolfram H. Dressler. 2020. "Forged in Flames: Indigeneity, Forest Fire, and Geographies of Blame in the Philippines." *Postcolonial Studies* 23, no. 4: 527–45.

Southern Adventist University. 2021. "Philippines—AFM: Palawano Project." Accessed March 4, 2024. https://www.southern.edu/administration/student-missions/calls/philippines-palawano-project.html.

Theriault, Noah. 2017. "A Forest of Dreams: Ontological Multiplicity and the Fantasies of Environmental Government in the Philippines." *Political Geography* 58: 114–27.

Theriault, Noah. 2019. "Unravelling the Strings Attached: Philippine Indigeneity in Law and Practice." *Journal of Southeast Asian Studies* 50, no. 1: 107–28.

Titeca, Kristof, Patrick Edmond, Gauthier Marchais, and Esther Marijnen. 2020. "Conservation as a Social Contract in a Violent Frontier: The Case of (Anti-) Poaching in Garamba National Park, Eastern DR Congo." *Political Geography* 78: 1–9.

Tsing, Anna. 1999. "Becoming a Tribal Elder, and Other Green Development Fantasies." In *Transforming the Indonesian Uplands: Marginality, Power, and Production*, edited by Tania Li, 159–202. Amsterdam: Harwood.

Tsing, Anna. 2005. *Friction: An Ethnography of Global Connection*. Princeton, NJ: Princeton University Press.

van Beek, Walter. 1990. "The Culture in-between: Anthropologist and Missionary as Partners." In *The Ambiguity of Rapprochement: Reflections of Anthropologists on Their Controversial Relationship with Missionaries*, edited by Roland Bonsen, Jelle Miedema, and Hans Marks, 101–20. Nijmegen: Focaal.

van Schendel, William. 2002. "Stateless in South Asia: The Making of the India-Bangladesh Enclaves." *Journal of Asian Studies* 61, no. 1: 115–47.

van Schendel, William. 2005. "Geographies of Knowing, Geographies of Ignorance: Jumping Scale in Southeast Asia." *Environment and Planning D: Society and Space* 20, no. 6: 647–68.

van Schendel, William, and Erik De Maaker. 2014. "Asian Borderlands: Introducing their Permeability, Strategic Uses, and Meanings." *Journal of Borderlands Studies* 29, no. 1: 3–9.

Vayda, Andrew P. "Progressive Contextualization: Methods for Research in Human Ecology." *Human Ecology* 11: 265–81.

Vidal y Soler, Sebastian. 1874. *Memoria sobre el Ramo de Montes en las Islas Filipinas*. Madrid: Aribau.

Vitug, Marites Dañguilan. 1993. *Power from the Forest: The Politics of Logging*. Manila: Philippine Center for Investigative Journalism.

Vitug, Martes. 2000. "Forest Policy and National Politics." In *Forest Policy and Politics in the Philippines: The Dynamics of Participatory Conservation*, edited by Peter Utting, 11–39. Quezon City: Ateneo de Manila University Press.

Warren, James. 2007. *The Sulu Zone, 1768–1898: The Dynamics of External Trade, Slavery, and Ethnicity in the Transformation of a Southeast Asian Maritime State*. Singapore: National University Press.

West, Paige. 2006. *Conservation Is Our Government Now*. Durham, NC: Duke University Press.

Wheatley, Jeffrey. 2018. "US Colonial Governance of Superstition and Fanaticism in the Philippines." *Method & Theory in the Study of Religion* 30, no. 1: 21–36.

Whitford, Harry Nichols. 1921. "Forests and Human Progress." *Journal of Forestry* 19, no. 1: 58–64.

Wilk, Richard. 1989. "Decision Making and Resource Flows within the Household: Beyond the Black Box." In *The Household Economy: Reconsidering the Domestic Mode of Production*, 23–52. London: Routledge.

Wilk, Richard. 1991. *Household Ecology: Economic Change and Domestic Life among the Kekchi Maya in Belize*. Tucson: University of Arizona Press.

Wilkins, Dominic. 2021. "Where Is Religion in Political Ecology?" *Progress in Human Geography*. 45, no. 2: 276–97.

Wolf, Diane. 1990. "Daughters, Decisions, and Domination: An Empirical and Conceptual Critique of Household Strategies." *Development and Change* 21, no. 1: 43–74.

Wolf, Diane. 1991. Female Autonomy, the Family, and Industrialization in Java." *Journal of Family Issues*. 9, no 1: 85–107.

Wolf, Eric. 1982. *Europe and the People without History*. Berkeley: University of California Press.

Worcester, Dean. 1901. *The Philippine Islands and Their People*. London: Macmillan.

Yusoff, Kathryn. 2017. "Indeterminate Subjects, Irreducible Worlds: Two Economies of Indeterminacy." *Body & Society* 23, no. 3: 75–101.

Zon, Raphael. 1920. "Forests and Human Progress." *Geographical Review* 9: 139–66.

Index

Acre, Friar Pedro, 84
adat, 34, 46
Adventist Frontier Missions (AFM), 103
  approach of, 5
  beljan as targeted by, 168–70, 187–88
  biopolitics of, 125–28, 160–62, 178–79,
    200n10
  complex near Brooke's Point, 204n6
  entrustment and coercion of, 187–88
  influence on Pala'wan livelihoods, 170–71
  in Kamantian, 13–15, 16*f*, 17*f*
  and mediating reform practices, 109
  online literature of, 21, 192n9
  pastoral power and approach of, 176–78, 184
  reflexivity of, 192n8
  reform practices adopted by, 158, 190
  resistance to, 175–76, 188–89
  rise, structure, and interventions of, 111
  ritual practices and use of customary objects
    prohibited by, 6, 130, 161, 162, 164–65,
    171–73
  sanctions broken strictures and norms, 173–75
  and study method, 21
  See also Seventh-day Adventists
afterlife, 195n9, 204n3. *See also* death
Agamben, Giorgio, 10–11
Agrawal, Arun, 192n1
agriculture
  and accumulation and concentration of
    wealth in urban areas, 156
  under American occupation, 89, 94, 97–98
  delineation of field boundaries, 196n13
  gender roles in, 62, 63, 196n16
  limitations placed on Pala'wan, 1–3
  priorities in, 120–21
  rattan, 197n22
  and residential patterns, 48–49
  rice as personhood, 53–57
  shifts to more intensified commercial,
    197n20
  under Spanish colonialism, 80, 82–84
  See also deforestation; forest conservation;
    Palawan Integrated Development Project
    (PIADP); Palawan Tropical Forest
    Protection Programme (PTFPP); swidden

Aguilar, Sofrano, 134, 135, 137, 138
Ahern, George, 94
Alvarez, Jose, 201n5
American occupation, 88–89, 108–10
  Baptist missionaries under, 100–103
  biopolitics of Christianity and tribal "Other"
    under, 98–100
  civilizing policies under, 96–97
  criminalization of swidden and optimization
    of tribal uplands under, 93–96
  legacies of, 7
  in-migration, land use, and Pala'wan
    marginalization under, 97–98
  postcolonial biopolitical continuities,
    103–8
  reforestation efforts under, 199n2
  tribal anticitizens under, 89–93
ancestral domains, 105–7
Ancestral Domain Sustainable Development
  and Protection Plans (ADSDPPs), 107
Andaya, Leonard, 199n7
Anderson, Eric, 80
animals, 51. *See also* food restrictions
anticitizens, 78, 89–93, 189–90
Anting-Anting, 35, 98–99
Apostolic Vicariate of Puerto Princesa,
  201n4
Appadurai, Arjun, 33
Aquino, Cory, 104–5, 109
assimilation
  and civilizing policies in southern Palawan
    under American occupation, 96–97
  and criminalization of swidden and
    optimization of tribal uplands, 93–96
  and 1903 census, 89–93

babalyan (Tagbanua), 34–35, 198n24
"backsliding," 173
Bagadion, Benjamin, 199n2
Bagrar, Sonja, 176
bagtik, 139
Baptist missionaries, 100–103
Barrows, David, 90, 93
basakan (rice paddy systems), 97–98
Bataraza, 18*map*